The applied mycology of *Fusarium*

The applied mycology of *Fusarium*

SYMPOSIUM OF
THE BRITISH MYCOLOGICAL SOCIETY
HELD AT QUEEN MARY COLLEGE
LONDON, SEPTEMBER 1982

EDITED BY
MAURICE O.MOSS & JOHN E.SMITH

CAMBRIDGE UNIVERSITY PRESS
CAMBRIDGE
LONDON   NEW YORK   NEW ROCHELLE
MELBOURNE   SYDNEY

CAMBRIDGE UNIVERSITY PRESS
Cambridge, New York, Melbourne, Madrid, Cape Town,
Singapore, São Paulo, Delhi, Tokyo, Mexico City

Cambridge University Press
The Edinburgh Building, Cambridge CB2 8RU, UK

Published in the United States of America by Cambridge University Press, New York

www.cambridge.org
Information on this title: www.cambridge.org/9780521279246

First published 1984
First paperback edition 2011

A catalogue record for this publication is available from the British Library

Library of Congress Catalogue Card Number: 83-5337

ISBN 978-0-521-25398-7 Hardback
ISBN 978-0-521-27924-6 Paperback

Contents

Contents

Contributors

C. Anderson, *RHM Research Ltd. The Lord Rank Research Centre, Lincoln Road, High Wycombe, Bucks. HP12 3QR, UK.*

P. K. C. Austwick, *Aerobiology Unit, Cardiothoracic Institute, Brompton Hospital, Frimley, Surrey GU16 5QE, UK.*

C. Booth, *Commonwealth Mycological Institute, Ferry Lane, Kew, Surrey TW9 3AF, UK.*

J. D. Bu'Lock, *Wolfson Biomass Unit, Weizmann Microbial Chemistry Laboratory, Department of Chemistry, The University of Manchester, Manchester M13 9PL, UK.*

J. H. Burnett, *University of Edinburgh, Old College, South Bridge, Edinburgh EH8 9YL, UK.*

M. L. C. Chiu, *Department of Bioscience and Biotechnology, Applied Microbiology Division, Royal College Building, 204 George Street, Glasgow G1 1XW, UK.*

N. Claydon, *Glasshouse Crops Research Institute, Worthing Road, Rustington, Littlehampton, West Sussex BN16 3PU, UK.*

R. B. Drysdale, *Department of Microbiology, The University of Birmingham, P.O. Box 363, Birmingham B15 2TT, UK.*

J. Gilbert, *Ministry of Agriculture, Fisheries and Food, Food Laboratory (Norwich), Haldin House, Old Bank of England Court, Queen Street, Norwich NR2 4SX, UK.*

J. F. Grove, *Agricultural Research Council Unit of Invertebrate Chemistry and Physiology, The University of Sussex, Brighton, Sussex BN1 9RQ, UK.*

R. Marchant, *School of Biological and Environmental Studies, New University of Ulster, Coleraine, Co. Londonderry BT52 1SA, Northern Ireland, UK.*

C. J. Mirocha, *Department of Plant Pathology, 304, Stakman Hall of Plant Pathology, 1519 Gortner Avenue, St. Paul, Minnesota 55108, USA.*

L. Mitchell, *Department of Bioscience and Biotechnology, Applied Microbiology Division, Royal College Building, 204 George Street, Glasgow G1 1XW, UK*

M. O. Moss, *Department of Microbiology, University of Surrey, Guildford, Surrey GU2 5XH, UK.*

D. Price, *Glasshouse Crops Research Institute, Worthing Road, Rustington, Littlehampton, West Sussex BN16 3PU, UK.*

J. E. Smith, *Department of Bioscience and Biotechnology, Applied Microbiology Division, Royal College Building, 204, George Street, Glasgow G1 1XW, UK.*

G. L. Solomons, *RHM Research Ltd, The Lord Rank Research Centre, Lincoln Road, High Wycombe, Bucks HP12 3QR, UK.*

J. L. Thomas, *Revlon Health Care (UK) Ltd, Station Road, Shalford, Surrey GU4 8HE, UK.*

Preface

The genus *Fusarium* contains a number of species of moulds which have been recognised for a long time as being important plant pathogens responsible for wilts, blights, root rots and cankers in a very wide range of important crop plants, including trees. Species of *Fusarium* are world-wide in their distribution and may be isolated from soil and decaying organic material, particularly of plant origin. More recently the genus has acquired notoriety because of the ability of several species to produce toxic metabolites causing illness and even death in man and his domesticated animals.

Members of the genus also play a role in the biodegradation of organic materials, in the post-harvest spoilage of crops, and in the biodegradation of, for example, pharmaceutical products. Some species are pathogens of insects and others are agents of human disease and possibly allergies.

Although these activities reflect adversely on the influence of the genus on the well-being of man, some aspects of the biology and biochemical versatility of *Fusarium* may be turned to benefit. Pathogenicity to insects may be used to control insect pests, a number of secondary metabolites are of commercial value, and the primary metabolism of strains of at least one species has been utilised in the production of microbial biomass which can be readily converted into a wide range of foodstuffs.

The diverse facets of this important genus of moulds formed the substance of a symposium on 'The Applied Mycology of *Fusarium*' organised by the Physiology Committee of the British Mycological Society and held at Queen Mary College, London during September 1982. This book contains the major contributions to that symposium following in the tradition of the first of the published symposia of the Society on the genetics and physiology of *Aspergillus*.

We thank all the contributors for providing the manuscripts and the staff of Cambridge University Press for their help with the layout and design of the book.

Maurice O. Moss
Department of Microbiology
University of Surrey
Guildford

John E. Smith
Department of
  Bioscience and Biotechnology
University of Strathclyde
Glasgow

1
The *Fusarium* problem: historical, economic and taxonomic aspects

C.BOOTH

Commonwealth Mycological Institute, Ferry Lane, Kew, Surrey TW9 3AF, U.K.

Historical aspects of the *Fusarium* problem

When did our knowledge of *Fusarium* begin? There is a tendency to regard it as all very new, in fact I recently read a thesis in which it was claimed that our knowledge of *Fusarium* diseases of cereals began with a paper published by Bennett in 1928.

I can assure you that it began much earlier; precisely when depends upon what you mean by the *Fusarium* problem. One of the first written descriptions of ear rot of maize caused by *F. moniliforme* was described from native Aztec descriptions in the sixteenth century by a Franciscan friar in Mexico. When the plant pathologists first went overseas to look at peasant agriculture they found the farmers had local names for the diseases that occurred in their crops, and these names obviously went back for generations, usually being translated as red mould, white mould etc.

If we mean when were *Fusarium* diseases first known within the concept of modern terminology in a scientific system, then of course the date is 1809, when Link first described the genus. He described *Fusarium roseum* as the first species; unfortunately his collection was mixed and it has been used as a mixed-up name ever since. It is however surprising in such a supposedly unstable genus and in an era when we, as taxonomists, are constantly accused of changing names, how many of the original names have survived unchanged: *F. lateritium* Nees described 1817; *F. heterosporum* Nees described 1818; *F. oxysporum* Schlechtendahl described 1824; *F. avenaceum* Fries (*Fusisporium*) described 1832; *F. equiseti* Corda (*Selenosporium*) described 1838; *F. graminearum* Schwabe described 1838; *F. solani* (*Fusisporium*) Martius described 1842; *F. sambucinum* Fuckel described 1869; *F. semitectum* Berk. & Rav. described 1875.

The *Fusarium* problem can be divided into four major aspects. These are: storage rots, plant pathogens, toxins and hormones, and human and animal pathogens.

Fusarium *as the cause of storage problems*

A study of the problems caused by *Fusarium* began in a modern sense with an investigation into the rotting of potatoes carried out by Martius in 1840–41 and published in 1842. He found the causal organism to be a fungus which he called *Fusisporium solani*; this was later transferred to *Fusarium* as *Fusarium solani* (Mart.) Sacc.

Throughout the mid-nineteenth century a number of papers appeared describing *Fusarium* species associated with rots of potato tubers. The workers were mostly German and included Harting (1846), Schacht (1856) and Reinke & Berthold who wrote a marvellous paper called Zersetzung der Kartoffel (the decay of potatoes), published in 1879. All these authors regarded the *Fusarium* species they isolated as saprophytes. In fact both de Bary (1861) and Reinke & Berthold (1879) regarded them as obligate saprophytes.

It was almost 20 years later that Pizzigoni (1896) and Wehmer (1897) showed by inoculation experiments that *Fusarium* species could cause tuber rots. These findings were not immediately accepted because they contradicted the authority of the great de Bary and of Reinke and Berthold and also because certain of their contemporaries such as Frank (1896 and 1898) tried and failed to produce comparable results. In fact it was not until 1904 that the parasitic nature of *Fusarium* species as storage rots was established in a paper published by Erwin F. Smith and Swingle (1904) in the U.S.A. Unfortunately Smith and Swingle regarded *F. solani* and *F. oxysporum* as identical species and took up *F. oxysporum* because it was the earlier name. Papers by Appel & Wollenweber (1910), Sherbakoff (1915) and others have since added a number of other *Fusarium* species to the list of storage rots.

Fusarium *as the cause of plant disease*

It is also to workers in the U.S.A. in the latter part of the last century that we owe our first specific knowledge of *Fusarium* species as the causal agents of serious plant diseases and also as toxin producers which could cause serious problems when infected grain was fed to animals.

Fusarium oxysporum featured as the first and it is still the most important *Fusarium* species causing diseases of economic crops. Between 1892 and 1899 a number of papers were published in the U.S.A. demonstrating the pathogenicity of *Fusarium oxysporum* to living plants. Two papers in particular stand out. A disease of cotton called 'Frenching' was becoming serious in Alabama, and Atkinson was invited to investigate it; in 1892 he published his findings and described *Fusarium vasinfectum* as the causal organism. This is a form of *F. oxysporum* and the name is still used in relation to cotton wilt. Atkinson described the typical *Fusarium* wilt including the presence of gummy substances blocking the vascular tissue, and he illustrated what has proved to be the diagnostic character for the identification of *F. oxysporum*, namely the phialides producing the microconidia.

The second outstanding paper was by Erwin F. Smith. During the 1890s he extended this work by Atkinson to the wilt disease of cotton, watermelon and cowpea in South Carolina and applied the technique of true pathogenicity testing, almost to the use of Koch's principles. He went on to tell us (1899) that a field infected with melon wilt fungus should not be planted with melons for a number of years but that canteloupes, cotton, peanuts, cowpeas or soybean could be grown. He also demonstrated how *Fusarium* was carried by seeds, water run-off, soil particles or farm implements. Smith called his fungi: *F. vasinfectum* (*Neocosmospora vasinfecta*) from cotton, *F. niveum* from water melon and *F. tracheiphila* from cowpea. He showed remarkable perception; he could not separate his three pathogenic fungi in pure culture but he used the three names because of their different pathogenicity.

It was not until 1940 that Snyder and Hansen placed these three names and about twenty other similar ones as synonyms of *F. oxysporum* and separated the pathogenic strains (because that is what they are) as formae speciales. With all due respect this was not a very remarkable advance after 40 years and now 40 years later we are still unable to differentiate one pathogenic strain from another without the help of a suitable host plant.

Fusarium *toxins*

Shortly after the work of Smith and his compatriots, evidence came to light regarding the serious toxic effects that may arise when grain infected with *Fusarium* species is fed to farm animals. In Nebraska in the 1890s, horses, cows and pigs were reported as losing hair and hooves after eating infected grain. The situation was summarised by Peters in a paper published in 1904. Much of this early work was confused because workers believed, as did Peters, that they were dealing with an outbreak of ergotism. However, by this time Sheldon (1904) had already confirmed that the toxicity was due to a *Fusarium* species which he isolated and named *F. moniliforme* – in fact the first *Fusarium* species to be named following an investigation into its toxicity.

Fusarium *infections of humans*

In recent years we have had many examples of *Fusarium solani* and *F. oxysporum* causing a direct fungal infection of the eye. Ironically these troubles only go back to around the time when antibiotics came into general use in the 1940s and to when, more specifically, the use of steroids began in the 1950s. There are also many records of *Fusarium* species associated with various types of ulcers. An interesting paper by Greco was published in 1916 on 'The origin of tumours'. He describes a fungal infection of the nose which he said was caused by *Fusarium vinosum*, a species Wollenweber placed as a synonym of *F. flocciferum*, one of the rarer members of the Discolor group.

What is *Fusarium* and what is its biology?

Originally, as proposed by Link in 1809, *Fusarium* was the name of hyaline, fusiform, non-septate asexual spores borne on a stroma. Note the 'non-septate'. A definition such as this is not specific and could apply to at least fifty different genera of Hyphomycetes. The 'borne on a stroma' was given considerable stress before the days of cultures and hence the genus was included by Fries (1821) in the Tuberculariaceae. With the development of pure culture methods for the identification of *Fusarium* species the presence of a stroma or sporodochium was no longer regarded as an essential character of the genus. For more details, see Wollenweber (1913).

Various people have tossed the generic description around in the past 70 years and all we can say is that *Fusarium* is basically a genus of hyaline, septate, phialidic (enteroblastic) asexual spores whose foot cell bears a heel. The only significant or specific character in this description is the heel. This whole genus is, if you like, bound together by a heel. The origin of the heel of the foot cell was considered by Wilcox *et al.* (1913) to be a function of the medium and growth rate, chiefly because it is absent from many conidia of the *Martiella* group, notably *Fusarium solani* and *F. coeruleum*. Pethybridge & Lafferty (1917) considered it to be a question of ageing as they found it to be more prominent on spores produced in older cultures. What is its biological or ecological function?

I would say nil. Yet obviously it binds together a group of species which have many specialised characters in common. In fact it transcends the divisions of the perfect state genera which have a *Fusarium* conidial state and it has made us look again at our classification of these perfect or teleomorphic states.

To me this heel on the basal cell of the macroconidium or asexual spore represents a vestigial character which illustrates a line of genealogical development of the highly organised *Fusarium* phialide from the more primitive sporogenous cell as found in *Trichothecium roseum*. More primitive forms are found in other conidial *Hypomyces* species. As the major toxins produced by *Fusarium* species are trichothecenes it is appropriate that there exists such a morphological link of this specialised nature between *Fusarium* and *Trichothecium*, from which the name for the toxin was derived.

The fact that *Fusarium* species have a highly effective enteroblastic phialidic method of spore formation implies that the conidia are slime spores and that they are dispersed by water, soil particles or on seed etc. The exception to this would be if the spore had undergone a secondary modification to dry spores as found in *Penicillium* and *Aspergillus* but there is no evidence of this in *Fusarium*. This statement is supported by the fact that although *Fusarium* species are very common, especially in soil, they are seldom caught in spore traps. Certainly in our *Fusarium* laboratory no *Fusarium* species were isolated from the air, nor do they normally occur as laboratory contaminants. However, there are exceptions. In the first place about half of the *Fusarium* species are capable of producing a perithecial state; records from aerial spore-trapping experiments seldom show whether a *Fusarium* colony developed from an ascospore or a conidium. Nelson, White & Toussoun (1971) demonstrated how effective the ascospores of *Gibberella zeae*, the ascospore stage of *F. graminearum*, could be in the dispersal of *Fusarium graminearum* in their carnation houses. The interesting point to me is the situation where *Fusarium* species without a known perithecial state have been authentically recorded from a spore trapping experiment. Such a report was produced by Lukezic & Kaiser (1966). They found that conidia of *Fusarium semitectum* (which they called *F. roseum* – Gibbosum), that formed on aerial mycelium, were easily dislodged by wind and dispersed in the atmosphere by wind speeds as low as 2.4 km.p.h. They also found that conidia formed on sporodochia on banana fruit could not be dislodged by air blasts; however, they offered no explanation of this apparent discrepancy. In fact the explanation is very simple: *Fusarium semitectum*, and several other species of *Fusarium*, have both slime-spored and often almost vestigial dry-spored conidia.

I think the precise method of spore formation in conidial fungi is a theme somewhat overplayed by taxonomists looking for genealogical relationships. There is a strict limit to the number of ways a conidium can be formed from a strand of mycelium and similar methods have obviously arisen several times in unrelated species. These various methods have been studied in great detail but what is biologically significant is whether a conidium ends up as a dry spore or a slime spore. In fact there is considerable evidence that many fungi produce both. *Fusarium* in the course of evolution has become adapted largely as a specialised soil fungus with the main spore slimy, but there is plenty of evidence available for the existence of a somewhat vestigial dry spore or blastosporic form. In *F. semitectum* and *F. avenaceum* these spores are roughly equal in size to the phialidic macroconidia whereas in *F. fusarioides* (*F. chlamydosporum*) and *F. sporotrichioides* the blastosporic form represents the microconidial state. It is the presence of these dry blastospore forms which explains frequent reports from aerial trapping experiments of *Fusarium* spores from species without a known perithecial state.

To return for the moment to my statement that the heel on the conidial cell transcends the perithecial state as a taxonomic character and makes one look again at our generic classification of these perithecial states. The Saccardoan emphasis on spore septation as a generic character within the Hypocreales has had to be discarded. As it existed, the three basic genera with *Fusarium* conidial states were *Nectria*, with yellow to red perithecia and 1-septate ascospores, *Calonectria*, with a similar pigmentation and 2- or 3-septate ascospores and *Gibberella*, also with 3-septate ascospores and a purple pigmentation in the perithecial wall. These are really very minor differences. Spore septation alone cannot be accepted as a generic character within the Hypocreales and for further discussion of this see recent papers by Booth (1978), Samuels (1978) and Amy Rossman (1983). In fact, ascospore septation is not a generic character and it is certainly a very minor diagnostic character compared to the heel on the foot cell of the phialidic macroconidium. One can say this, because this heel binds together a group of fungi which have numerous growth, pathogenic and toxigenic characters in common.

Fungi which have been placed in other genera, or even in another class, because of some stromatic character, but which produce septate conidia with a foot cell have been shown, in culture, to be typical *Fusarium* species. For instance, the Coelomycete genus *Botryocrea*, has a pycnidium-like structure producing *Fusarium*-like spores with a foot cell. This is merely an environmental reaction to dry conditions. In nature many stromatic fungi develop, when in a dry atmosphere, the first conidia below the upper surface of the stroma. *Botryocrea sclerotioides* develops its stromata on species of *Astragalus*. However, as Amy Rossman (1983) has just demonstrated, when this species is grown in culture it assumes normal *Fusarium*-like growth. *Pycnofusarium rusci* on *Ruscus aculeatus*, another pycnidial genus with a *Fusarium*-like spore and which is based on a single collection, will, I am prepared to wager, prove to be a similar case.

Speciation or species concepts

In this area it is my opinion that we largely create our own problems. *Fusarium* species are a labile group of organisms which when grown on artificial media may produce a multitude of variants both in their morphology and cultural characteristics depending upon the cultural conditions. There is no mystery about *Fusarium* taxonomy. In fact I can state categorically that if you make fresh collections from nature and place single conidium or single ascospore isolations on to a neutral culture medium then with the possible exception of *Fusarium avenaceum* you will find no greater variation in your resultant cultures than you would find in any other group of microfungi in culture. Of course if you are looking for trouble *Fusarium* cultures are ideal candidates for creating problems. William Brown, in a series of papers published between 1924 and 1928 (two of which were written with A. S. Horne), demonstrated how cultural conditions could affect both the morphology of the spores produced and the pigmentation of the medium. The C/N ratio and the amount and type of phosphate were all shown to be important factors.

Apart from the medium, if you are careless with your culture techniques you will produce your own problems, e.g. mass transfer from old cultures on to different media will often make you think you have contaminants present. In *Fusarium avenaceum*, probably the most unstable *Fusarium* species, mass transfer will soon reduce the long aciculate conidia to a short, variable almost microconidia-like mass; however, this is a cultural effect and if single conidia are taken then the normal form is quickly restored. Even preparing slides needs care. The tip of a needle introduced into a sporodochium will produce a slide with a spore morphology different from one made from the edge of a growing colony.

What we are aiming for is the concept that members of a species taxon are similar because they share a common heritage; they are not placed together because under certain conditions they can be made to look alike.

How do we define a species?

Alan Burgess (1955) wrote an interesting essay entitled 'problems associated with the species concept in mycology'. He compared species concepts for fungi with those for flowering plants and came to the conclusion that in flowering plants species concepts are extremely easy to formulate compared to those for fungi. It is interesting to note that whereas there are far more fungi then flowering plants, there are probably five times more flowering plant taxonomists than fungal taxonomists.

However, to return to fungal taxonomy. In spite of my categorical statement, variation does occur in *Fusarium* species and these variations often blur the characters one accepts for species limitation. What these limitations are depends upon a number of things and it is a question of determining the species characters within the limits of this natural variation. If the species has a known perithecial state, these limitations cover the range of variation that comes from a series of ascospore isolations on a range of what has come to be accepted as standard *Fusarium* media. One would not include the effects of staling on the spore morphology.

Some names are used to cover what at first appears to be a complex of species. In *Fusarium oxysporum* for instance, there are six or seven distinct morphological and cultural forms, all of which have in the past been given different names. A taxonomist would be happy to accept these as distinct species except that in doing so he would ignore the biology of the species. In *F. oxysporum* it is the strain of the organism that matters, i.e. its capacity to attack a specific host. We refer to these strains as formae speciales and the specific gene or genes they carry can be present in any one of the six or seven cultural forms mentioned.

There have been a number of myths perpetrated about the ability of *Fusarium* species to produce genetically induced variation. To some extent one must blame those who write reviews. Although there are people who are masters of this art, my complaint is that the reviewer likes to tell a story and tends to accept published statements which support it (in this case that of genetically controlled variation), instead of attributing so much of this apparent variation to slap-dash cultural methods.

One such story concerns the variation in *Fusarium* species due to heterokaryosis as a result of either branching of, or conidial production from, a heterothallic thallus. This has been maintained in a number of publications and is based on the most flimsy evidence. As Puhalla (1981) stated (and our own observations confirm), hyphal tips are uninucleate and both micro- and macroconidia begin with only one nucleus; thus all nuclei in a macroconidium are mitotic descendants of this original haploid nucleus. Certainly hyphal fusion occurs in cultures but I would suggest that this relates more to nutritional needs than to genetical exchanges. In the first place there is a tendency for one strain of a species to rapidly occupy a whole area, with fusion thus occurring between hyphae with the same mitotic history. Also the occurrence of new strains of formae speciales with their specific pathogenicity is an extremely rare phenomenon which would surely not be the case if genetic interchange was as frequent as the presence of hyphal fusion might suggest. Available evidence suggests that the spread of pathogenic strains outside their original or immediate area is due more to the ineptitude of man than to a new mutation. The occurrence of widely dispersed opposite strains of heterothallic species will be discussed later.

Another chestnut is parasexuality which is often quoted as a means by which *Fusarium* species achieve variability. However there is little supporting evidence and a clear demonstration of parasexuality in *Fusarium* species is still not available. Evidence suggests that true variation in *Fusarium*, when it does occur, is in fact due to mutational changes in the nuclei and there are various techniques by which these may be induced in nature. These are, in relationship to the growth rate of these fungi, extremely rare.

Fusarium sexual stages

Finally we need to discuss the perithecial or sexual stages of *Fusarium* species. I have said elsewhere that taxonomists are apparently more interested in sex than are species of *Fusarium*. Whilst we are struggling to find the sexual or perithecial state, and to study the sexual processes which produce them, *Fusarium* species are as rapidly as possible dispensing with these processes altogether. In the genus as a

whole those species which apparently live a life without sexual conjugation appear to compete equally with those that still have a sexual stage
in their life cycle. In fact *Fusarium oxysporum*, the most common
species, the most economically important and the most studied species,
has no perithecial state. Even in cold countries, where there may be up
to six months of enforced dormancy each year, there is no greater
predominance of species producing perithecia than those without this
facility. In fact, in contrast to most Pyrenomycetes, perithecial states of
Fusarium species which include *Nectria, Calonectria* and *Gibberella*
have a comparatively short period of survival, especially if dried. In
species with a sexual phase there is a tendency to change from a
homothallic to a heterothallic state and for one strain to dominate a
given area. *Fusarium solani*, a very common world-wide species, occurs
both as homothallic and heterothallic strains which produce perithecia
of *Nectria haematococca*. Each year the Commonwealth Mycological
Institute (CMI) receives from all over the world what may be regarded
as a random sample of isolates and the number of heterothallic isolates
amongst these is probably 100 for each homothallic isolate received. A
similar situation exists in *Fusarium graminearum (Gibberella zeae)* but
in this species homothallic isolates are more common. In *F. decemcellulare (Calonectria rigidiuscula)* a more specialised situation exists. This
tropical species has a more restricted distribution and, whereas in
homothallic strains of, for example, *F. solani* it is difficult to stop
perithecia being formed in culture, in *F. decemcellulare* perithecial
production has to be encouraged by specific cultural techniques. Also it
has been stated that strains of *C. rigidiuscula* with four-spored asci are
saprophytic whereas those with eight spores are parasitic. We have
observed only strains with four-spored asci to be homothallic.

Thus homothallic strains are definitely in the minority and our
evidence for stating that in heterothallic species asexual conjugation is
restricted comes from the observation that opposite mating types of
heterothallic strains are often widely dispersed geographically. Gordon
(1954) was probably the first worker to observe this phenomenon. He
found that the mating types of *Fusarium sulphureum (Gibberella
cyanogena)* occurred separately in nature and in no instance in his
experience were they found together. Thus mating type 'a' from British
Columbia mated with 'A' from Manitoba and Tasmania. Both were
found in Britain and in Prince Edward Island but not together in the
same area. Gordon went on to develop this work with 13 other
heterothallic species. He found that most species were bisexual but

self-sterile and opposite mating types were geographically separate, which explains the rarity of *Gibberella* species in nature. One is left with the impression that like galaxies these opposite mating strains are rushing away from each other.

Burnett (1975), in his review of the situation existing for heterothallic strains of *Nectria haematococca* f. sp. *cucurbitae*, stated that in nature no two morphological and physiological compatible types of species exist together. Now this may be true to a large extent for other species of *Fusarium* or at least true for *N. haematococca* f. sp. *cucurbitae*; however it is a fact that one does occasionally find perithecia of these heterothallic strains in nature, which suggests that these opposite strains do occasionally meet and overcome their apparent sexual aversion. Indeed, even with *F. sulphureum* Gordon did finally find one collection of *Gibberella cyanogena* in nature and we have collections of heterothallic *F. sambucinum* (*G. pulicaris*), *F. graminearum* (*G. zeae*) and several of *F. lateritium* (*G. baccata*), all of which provides evidence for heterothallic strains making contact in nature. Nevertheless the phenomenon is much rarer than one would expect. For the taxonomist this rarity does provide problems in establishing species limitations.

Having discussed the history and importance of *Fusarium* species I would like to end with a word of warning. If you set your sights on *Fusarium* species alone, and this is the subject of the Symposium, then you may be blinkered as to what is happening in related genera. Available evidence suggests that the production of trichothecenes is a phenomenon of the Hypocreales, the name itself being derived from *Trichothecium roseum*; therefore if you are concerned with these toxins you cannot be concerned with *Fusarium* alone. In plant disease, apart from the rather specialised 'oxysporum wilt' and the relationships which exist between *Fusarium* species and cereals, almost all symptoms of *Fusarium* diseases can be simulated by species of *Cylindrocarpon*, especially on acid soils. Also with regard to storage problems, other hypocreaceous fungi, such as *Trichoderma* and *Cylindrocarpon*, can cause the same problems and produce the same or similar toxins. So BEWARE.

References

Appel, O. & Wollenweber, H. W. (1910). Grundlagen einer Monographie der Gattung *Fusarium* (Link.). *Arbeiten aus der Kaiserlichen Biologischen Anstalt für Land- und Forstwirtschaft*, **8**(1), 1–207.

Atkinson, G. F. (1892). Some diseases of cotton. III. Frenching. *Bulletin Alabama Agriculture Experiment Station*, **41**, 19–29.

Bennett, F. T. (1928). On two species of *Fusarium*, F. culmorum (W.G.Sm.) Sacc. and F. avenaceum (Fries) Sacc., as parasites of cereals. *Annals of Applied Biology*, **15**, 213–44.

Booth, C. (1978). Presidential address. Do you believe in genera? *Transactions of the British Mycological Society*, **71**(1), 1–9.

Brown, W. (1925). Studies in the genus *Fusarium*. II. An analysis of factors which determine the growth-forms of certain strains. *Annals of Botany*, **39**(154), 373–408.

Brown, W. (1928). Studies in the genus *Fusarium*. VI. General description of strains, together with a discussion of the principles at present adopted in the classification of *Fusarium*. *Annals of Botany*, **42**(165), 285–304.

Brown, W. & Horne, A. S. (1924). Studies of the genus *Fusarium*. I. General account. *Annals of Botany*, **38**(150), 379–83.

Brown, W. & Horne, A. S. (1926). Studies in the genus *Fusarium*. III. An analysis of factors which determine certain microscopic features of *Fusarium* strains. *Annals of Botany*, **40**(157), 203–21.

Burgess, A. (1955). Problems associated with the species concept in mycology. In *Species Studies in the British Flora*, ed. J. E. Lousley, pp. 65–82. Arbroath, Scotland: T. Buncle & Co. Ltd.

Burnett, J. H. (1975). *Mycogenetics: An Introduction to the General Genetics of Fungi*. London: John Wiley.

De Bary, A. (1861). *Die gegenwärtig herrschende Kartoffelkrankheit, ihre Ursache und ihre Verhütung*. Leipzig: A. Förstner.

Frank, A. B. (1896). *Die Krankheiten der Pflanzen*, II. Breslau.

Frank, A. B. (1898). Untersuchungen über die verschieden Erreger der Kartoffelfäule. Die *Fusarium*-Fäule. *Bericht der Deutschen Botanischen Gesellschaft*, **16**, 279–80.

Fries, E. (1821). *Systema Mycologicum*, 1, XLI (intro.)

Gordon, W. L. (1954). Geographical distribution of mating types in *Gibberella cyanogena* (Desm.) Sacc. *Nature*, **173**, 505.

Greco, N. V. (1916). *Origine des Tumeurs (Etiologie du Cancer, etc.) et Observations de Mycoses (Blastomycoses, etc.) Argentines*. Buenos Aires: 'La Semana Medica', E. Spinelli.

Harting, P. (1846). Recherches sur la nature et les causes de la maladie des pommes de terre en 1845. *Nieuwe Verhandelingen de eerste Klasse van het Kon. Nederl. Inst. Van Wetensch*. Letterkunde en Schoone Künsten Amsterdam, XII, 203–97.

Link, J. H. F. (1809). *Observations in Ordines Plantarum naturales*, 1, 10. Berlin.

Lukezic, F. L. & Kaiser, W. J. (1966). Aerobiology of *Fusarium roseum* 'Gibbosum' associated with crown rot of boxed bananas. *Phytopathology*, **56**, 545–8.

Martius, C. F. P. von (1842). *Die Kartoffel-Epidemie der letzten Jahre oder die Stockfäule und Räude der Kartoffeln, geschildert und in ihren ursachlichen Verhaltnissen erörtert*. München: Akademie der Wissenschaften.

Nelson, P. E., White, B. L. & Toussoun, T. A. (1971). Occurrence of perithecia of *Gibberella* sp. on carnation. *Phytopathology*, **61**(6), 743–4.

Peters, A. T. (1904). *A Fungus Disease in Corn*. Seventeenth Annual Report of the Agricultural Experiment Station of Nebraska, pp. 13–18. Lincoln, Nebraska: The University of Nebraska.

Pethybridge, G. H. & Lafferty, H. A. (1917). Further observations on the cause of the common dry-rot of the potato tuber in the British Isles. *Scientific Proceedings of the Royal Dublin Society*, **15**, (NS), **21**, 193–222.

Pizzigoni, A. (1896). Cancrena secca ed umida della patate. *Nuovo giornale botanico italiano*. NS, **3**, fasc. 1, 50–53.

Puhalla, J. E. (1981). Genetic considerations of the genus *Fusarium*. In *Fusarium, Diseases, Biology and Taxonomy*, ed. P. E. Nelson, T. A. Toussoun & R. J. Cook, pp. 291–305. University Park & London: The Pennsylvania State University Press.

Reinke, J. & Berthold, G. (1879). *Die Zersetzung der Kartoffel durch Pilze*. Untersuchungen aus dem Botanischen Laboratorium der Universität Göttingen, I. Berlin: Wiegandt, Hempel & Parey.

Rossman, A. (1983). The phragmosporous species of *Nectria* and related genera. *Mycological Paper*, **150**. Kew: Commonwealth Mycological Institute.

Samuels, G. J. (1978). Some species of *Nectria* having *Cylindrocarpon* imperfect states. *New Zealand Journal of Botany*, **16**, 73–82.

Schacht, H. (1856). *Bericht an das Königl. Landes-Ökonomie-Kellegium über die Kartoffelpflanze und deren Krankheiten*. Berlin.

Sheldon, J. L. (1904). *A Corn Mould*. Seventeenth Annual Report of the Agricultural Experiment Station of Nebraska, pp. 23–32. Lincoln, Nebraska: The University of Nebraska.

Sherbakoff, C. D. (1915). *Fusaria of Potatoes*. Cornell University Agricultural Experiment Station Memoir No. 6. Ithaca, New York: University Press.

Smith, E. F. (1899). *Wilt Disease of Cotton, Watermelon and Cowpea*. US Department of Agriculture Bulletin No. 17. Washington: Government Printing Office.

Smith, E. F. & Swingle, D. B. (1904). The dry rot of potatoes due to *Fusarium oxysporum*. *USDA Bureau of Plant Industry*, Bulletin No. 55. Washington: Government Printing Office.

Snyder, W. C. & Hansen, H. N. (1940). The species concept in *Fusarium*. *American Journal of Botany*, **27**(2), 64–7.

Wehmer, C. (189,7). Untersuchungen über Kartoffelkrankheiten II. Ansteckungsversuche mit *Fusarium solani*. *Zentralblatt für Bakteriologie, Parasitenkunde und Infektionskrankheiten*, Abt. II, 3(25/26), 727–42.

Wilcox, E. M., Link, G. K. K. & Pool, V. W. (1913) *A Dry-Rot of the Irish Potato Tuber*. Research Bulletin No. 1. Nebraska Agricultural Experiment Station.

Wollenweber, H. W. (1913). Studies on the *Fusarium* problem. *Journal of Phytopathology*, **3**(1), 24–50.

2
The ultrastructure and physiology of sporulation in *Fusarium*

R.MARCHANT

School of Biological and Environmental Studies, New University of Ulster, Coleraine, County Londonderry BT52 1SA, Northern Ireland, U.K.

Introduction

It is no doubt due to the immense impact of members of the genus *Fusarium* on the activities of man that we find enormous numbers of papers in the literature covering all aspects of the biology of these ubiquitous organisms. However, when we come to consider some specific aspects of this interesting group of fungi we often discover that this plethora of published information contains little of substantive value. If we wish to discover basic information concerning the biology of sporulation in *Fusarium* we immediately encounter this problem of a large potential pool of knowledge which in reality is only a shallow puddle. Much of what has been written about sporulation in *Fusarium* is only concerned with the observed phenomenon of these organisms in the field and very little deals with the basic problems of spore ontogeny and the physiological processes leading to the production of spores. In this review I will try to pick out those data which appear to me to be relevant to an understanding of all the processes which lead to the formation of spores in *Fusarium* and also to indicate those areas where our knowledge is sadly lacking.

As with most groups of fungi the form and mode of production of spores plays a central role in the classification and identification of members of the group. In the fusaria the whole problem is compounded by the fact that the spores we recognise from the genus only represent the anamorphs which are associated with teleomorphs from a number of different genera within the Hypocreales. If we have little information on the ultrastructure and physiology of sporulation of the anamorphs then it is probably true to say that we have even less on the teleomorphs. This treatment of the material will, therefore, for several good reasons. be confined to the spore forms recognised as fusaria.

There are three clearly defined spore types which we can identify in the many species of *Fusarium* which have been named since the genus was erected by Link in 1809. The predominant spore type is the macroconidium which may co-exist with the microconidium in some species. The third spore form is the chlamydospore which can be formed from a hyphal starting point or from a conidial precursor. The bulk of investigation has necessarily been directed towards the macroconidium, although in recent years more attention has been turned towards the chlamydospore, probably because of its assumed and later proven importance in the survival of fusaria in the environment (Sitton & Cook, 1981).

It was assumed for a considerable time that the sporogenous cell which leads to the production of conidia was a phialide (Mason, 1933), however, blastic conidiogenous cells have been illustrated by Seemüller (1968) and Messiaen & Cassini (1968) cited by Booth (1971). It is hardly surprising that the genus should contain a number of different modes of conidium ontogeny in view of the diversity of teleomorph states associated with it. This variation within the genus has not been exploited to any great extent in the taxonomic studies nor has it been fully investigated from the structural and ontogenic viewpoint.

The ultrastructure of spores

The majority of investigations of the ultrastructure of *Fusarium* spores have been carried out on macroconidia of only a few species. The first of these studies were on the macroconidia of *Fusarium culmorum* (Marchant, 1966*a, b*) and used a relatively poor fixation procedure with potassium permanganate followed by embedding in epoxy resin. Nevertheless the essential features of the spore ultrastructure could be clearly discerned. The multiseptate spores were shown to have a nucleus in each cell of the conidium (Fig. 2.1) and, apart from the normal complement of organelles, to contain a substantial and obvious lipid content in the form of discrete droplets around the periphery of each cell (Fig. 2.2). This lipid appears to be a storage material used during the germination of the conidium, a fact demonstrated by Mumford & Pappelis (1978) using the technique of quantitative interference microscopy. Additional features demonstrated in these first investigations were the presence of single septal pores in the centre of each conidial septum. These pores have gently tapered margins, typical of ascomyce-

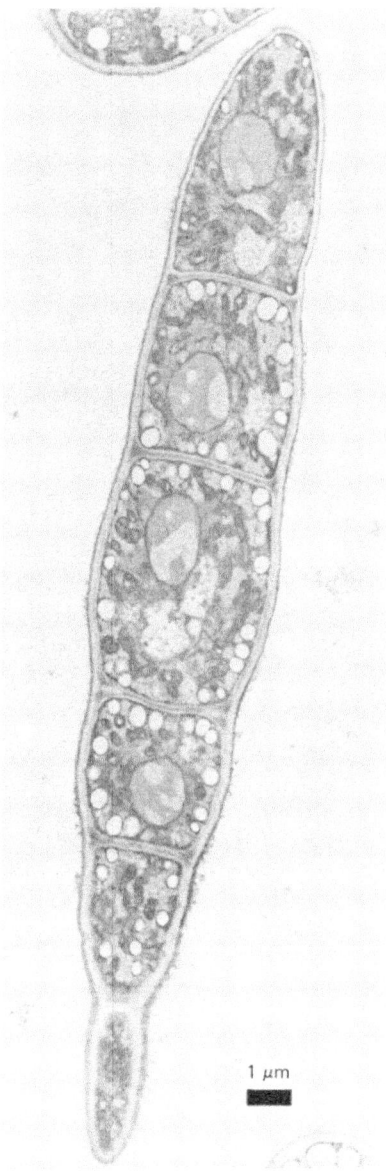

Fig. 2.1. Longitudinal section through a macroconidium of *Fusarium culmorum*, showing nuclei in each cell, mitochondria, vacuoles and small amounts of lipid.

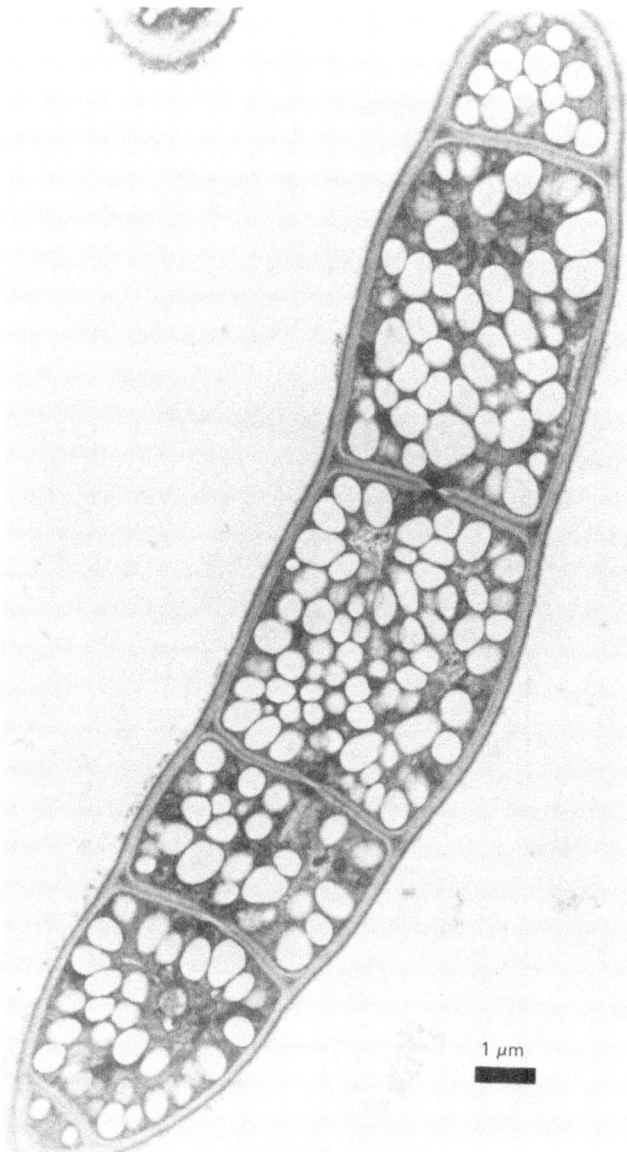

Fig. 2.2. Longitudinal section through a macroconidium of *Fusarium culmorum*, showing extensive storage lipid and perforate conidial septa with plugs.

tous fungi, and also have several associated Woronin bodies on each side of the septum. The development of Woronin bodies from microbodies has been shown from a study of the microconidia of *Fusarium oxysporum* f. sp. *lycopersici* by Wergin (1973) in an almost unique paper on the structure of these interesting organelles. During the prematuration stages of the macroconidia the Woronin bodies clearly lie to either side of the septum but later one appears to form a plug in the septal aperture such that the cells of the conidium become isolated from each other (Fig. 2.2) (Marchant, 1966a). Since the appearance of the initial studies on *F. culmorum* two further papers have appeared. García Acha *et al.* (1966) deduced from some rather poor micrographs that the conidial cells contained one or two nuclei and that the conidial septa did not have pores; however, a more recent paper published in Czech confirms the presence of septal pores in *F. culmorum* macroconidia (Šrobárová & Šrobár, 1979). Essentially similar structural features to those in *F. culmorum* have been observed in macroconidia of *F. solani* by Tewari & Skoropad (1975) using improved fixation methods for transmission electron microscopy and additionally employing freeze-etching and scanning electron microscopy. They showed single septal perforations in the conidia, Woronin bodies and septal plugs and large quantities of storage lipid. A brief paper in 1969 (Akai, Fukutomi & Kobayashi, 1969) had shown some similar features for macroconidia of *F. oxysporum* f. *cucumerinum* including the presence of only a single nucleus in each cell of the spore. The interest in the presence or otherwise of septal perforations and the form of these perforations stems from a paper describing the presence of multiperforate septations in hyphae of *F. solani* f. *phaseoli* (Reichle & Alexander, 1965), although even here the conidia are described with uniperforate septa.

A further area of disagreement in the literature has centred around the structure and arrangement of the surface layers of the conidia. The chemical and substructural features of the walls will be dealt with later, but in sectioned material various authors have discussed the numbers of layers observed within the walls. As I have suggested previously (Marchant, 1979), the image which one obtains in sectioned wall material depends to a large extent on the preparative methods used and in any case work on various cell walls of fungi has shown that it is unrealistic to expect the various components constituting the wall to be separated into totally discrete layers (van der Valk, Marchant & Wessels, 1977). Nevertheless two layers have been reported for the walls of *F. sulphureum* (Schneider & Seaman, 1974b; Schneider *et al.*,

1977) and *F. culmorum* (Campbell & Griffiths, 1974); three for *F. culmorum* (Marchant, 1966*a*; Laborda *et al.* 1974) and four for *F. solani* (van Eck & Schippers, 1976). A further outer layer of mucilage was described by Marchant (1966*a*) for the conidia of *F. culmorum* (see Figs 2.3 and 2.4), an observation which was later challenged by some workers (Akai, Fukutomi & Kobayashi, 1969) but confirmed by others (Tewari & Skoropad, 1975; Kleinschuster & Baker, 1974). It seems likely that this apparent inconsistency arises from the different modes of spore production which are possible even within the same species or within the same strain grown under different conditions. Booth (1971) explains how conidia may be produced either on aerial mycelium or produced as pionnotes which give the culture a flat greasy appearance. Even this may not provide the complete explanation, however, because conidia produced in submerged liquid culture can also show evidence of additional material outside the normal discrete wall layers (Marchant, 1975).

Many fusaria produce chlamydospores either within the hyphae or from individual cells of the conidia. Very many environmental conditions seem to favour the production of such spores which are generally taken to be resistant structures produced in response to adverse conditions. Examples of the range of conditions which have been reported to favour chlamydospore formation are starvation (Schneider & Seaman, 1974*a*); various nutritional requirements (Barran, Schneider & Seaman, 1977; Griffin, 1976), low pH (Griffin, 1976) and soil microorganisms or soil constituents (Ford, Gold & Snyder, 1970*a*, *b*). The physiological and metabolic interactions which lead to the formation of chlamydospores are therefore likely to be complex and have not yet attracted the level of investigation necessary to yield satisfactory explanations.

In relation to macro- and microconidia the structure of chlamydospores differs principally in the presence of a thickened wall and some simplification of the internal organisation. The stages of wall development will be dealt with in the following section on spore ontogeny. In common with the macroconidia both hyphal and conidial chlamydospores contain large quantities of lipid, together with a nucleus in each spore, an endoplasmic reticulum system and a number of apparently functional mitochondria. Apart from the obvious elaboration of wall structure there are therefore no particularly distinctive structural features of chlamydospores which separate them from other spore types (Stevenson & Becker, 1972; Schneider & Seaman, 1974*b*).

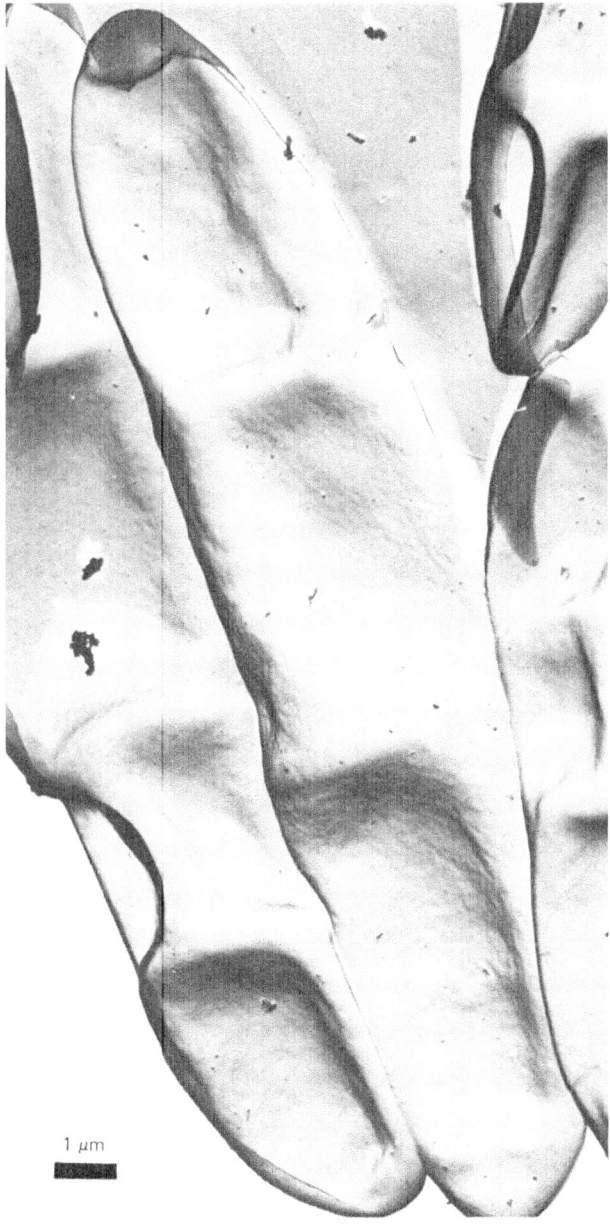

1 μm

Fig. 2.3. Replica of ungerminated macroconidia of *Fusarium culmorum* illustrating the mucilage layer which obscures detail of the underlying wall structure.

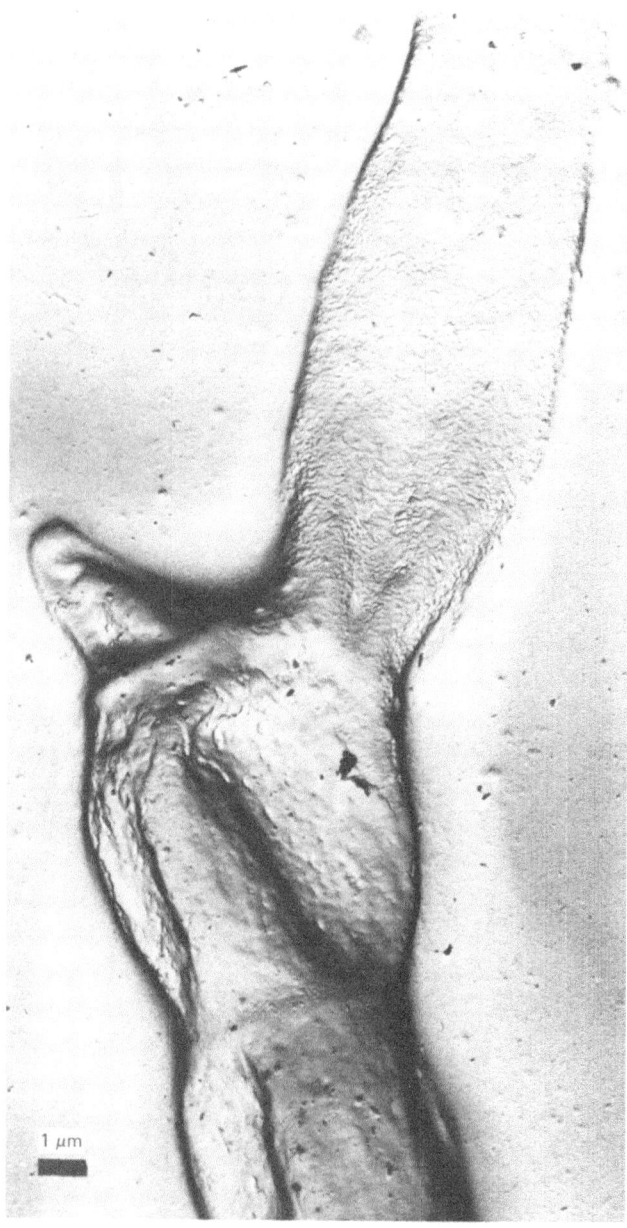

Fig. 2.4. Replica of a germinating macroconidium of *Fusarium culmorum*, showing the presence of mucilage on the conidium but not on the emerging germ tube.

The ontogeny of spores

In the recent book entitled 'Patterns of development in conidial fungi' (Cole & Samson, 1979) only scant reference is made to the development of spores in fusaria, which is purely a reflection of the small amount of basic work carried out in this area.

The development of macroconidia

The macroconidia have largely been considered to be produced from phialidic conidiogenous cells and diagrams of conidia in publications generally assume this point. Some indication of the possible mode of spore formation has come from Subramanian (1971), who, in a chapter devoted to the phialide, illustrates the formation of phialospores in *F. decemcellulare*. The essential feature of this phialidic sequence is that the spore forms within an enveloping structure and only subsequently erupts from it. This is in marked contrast to other blastic modes of formation where the spore is initiated by an extension of existing wall layer(s) and then continues to expand and mature. The phialidic system may give rise to only a single conidium from the phialide or may produce a succession of conidia. In the second instance the first produced conidium has a slightly different ontogeny from the succeeding ones. The illustration of Subramanian (1971) shows a primary conidium formed from a phialide where the original enveloping wall becomes ruptured to leave a collar round the tip of the phialide and a cap of wall material over the tip of the developing conidium. Further evidence that this development sequence occurs in other species of *Fusarium* is provided by the immuno-fluorescence work of Goos & Summers (1964) using *F. oxysporum* f. *cubense*. They were able to demonstrate a brightly fluorescent cap over the tip of the conidium which correlated with the similarly fluorescent collar of the phialide tip. The collar around the phialide can easily be shown by scanning electron microscopy although the cap over the conidium, being less persistent, cannot be demonstrated so easily.

Considering the number of anamorphic forms of fungi in which spore ontogeny has been fully investigated at the ultrastructural level, it is perhaps surprising to find only two papers in the literature dealing with macroconidium development in *Fusarium*. It is always difficult to put together a temporal development sequence from a random sampling method. The problems can be greatly eased by utilising either a highly selective system for each stage or to use some mechanisms for synchronising development. It was this second alternative which was used to obtain material for an ultrastructural investigation of phialospore formation in *F. culmorum* (Marchant, 1975). The physiological basis for this system was derived from the work of Larmour & Marchant (1977*a*) using cultures grown in a chemostat. As will be explained more fully later a stable non-sporulating culture can be switched to conidial production simply by increasing and holding the pH of the culture at 6.5 instead of 3.5. Samples of material taken over the ensuing 24 hours provide all stages of macroconidium development.

The scheme of conidiogenesis reported for *F. culmorum* although superficially phialidic in nature did show some features which indicated some thallic tendencies. The first recognisable stage was the production, within an apex, of a membrane bound region delimited by the fusion of vesicles (Fig. 2.5). This region contained a single nucleus and further development led to the deposition of wall material (the future wall of the conidium) around the outside of the internal membrane. Deposition of all material in this manner differs from the other major internal spore generation system studied in fungi, the ascospore, where wall material is deposited between a pair of membranes (Carroll, 1969). The cytoplasm excluded from the developing conidium degenerates to leave the incipient conidium with a forming septum at its base, complete with Woronin bodies (Fig. 2.6), which later becomes the abscissional region for the conidium. Further growth of the conidium leads to the rupture of the containing wall, to form the phialide collarette (Fig. 2.7) and the cap of wall material over the conidium. The final stages of conidium maturation involve the differentiation of the foot region, nuclear division and septation of the conidium (Fig. 2.8). This description covers only the formation of a first conidium and indeed does not indicate whether subsequent conidia could be produced from a single phialide; however, further conidia could be produced in a similar manner. The main reason for suggesting that this may not be a completely typical phialide system is that subsequent conidia could not be produced in basipetal sequence from a fixed conidiogenous locus as defined in the 'Proceedings of the First International Conference on Taxonomy of Fungi Imperfecti' (Kendrick, 1971). Although this study (Marchant, 1975) provides a basis for further investigation of conidiogenesis in fusaria it certainly does not provide all the information we would like to have even about *F. culmorum*. In particular the process of conidial septation has not been followed closely, although in fairness septation is not adequately described in the ascomycetous fungus system. The nuclear division in the spore requires further study and it

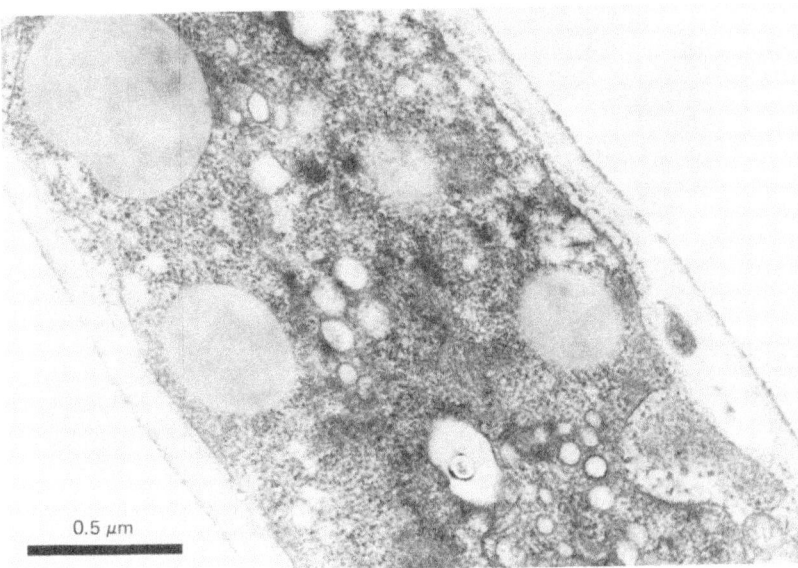

Fig. 2.5. Section through a developing phialide of *Fusarium culmorum*, showing the vesicles which fuse to produce the spore plasmalemma and subsequently contribute the developing spore wall material.

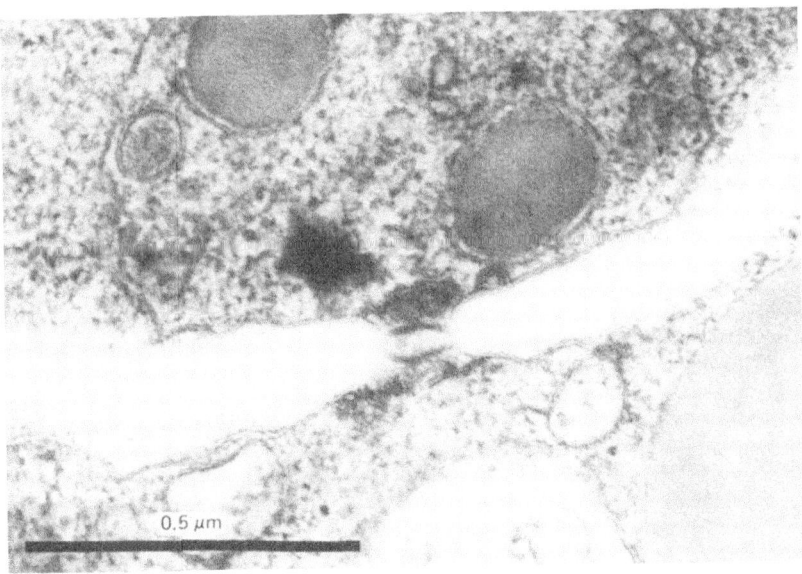

Fig. 2.6. Section through a developing septum at the base of a conidium in *Fusarium culmorum*. Note the presence of Woronin bodies near the septum.

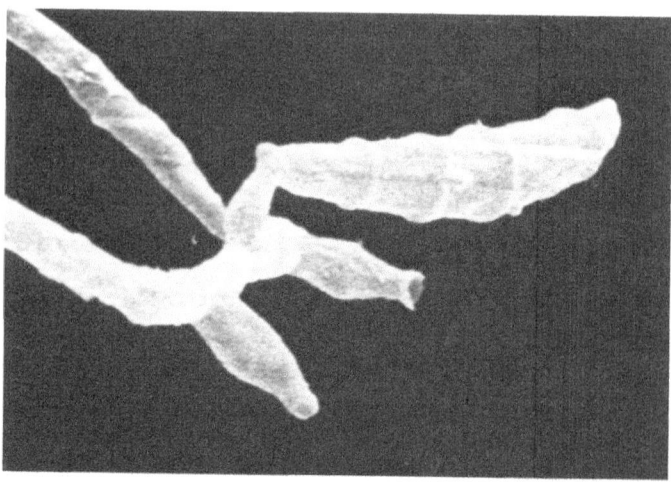

Fig. 2.7. Scanning electron micrograph showing the collarettes around phialide tips from which macroconidia have been released in *Fusarium culmorum*. × 1800.

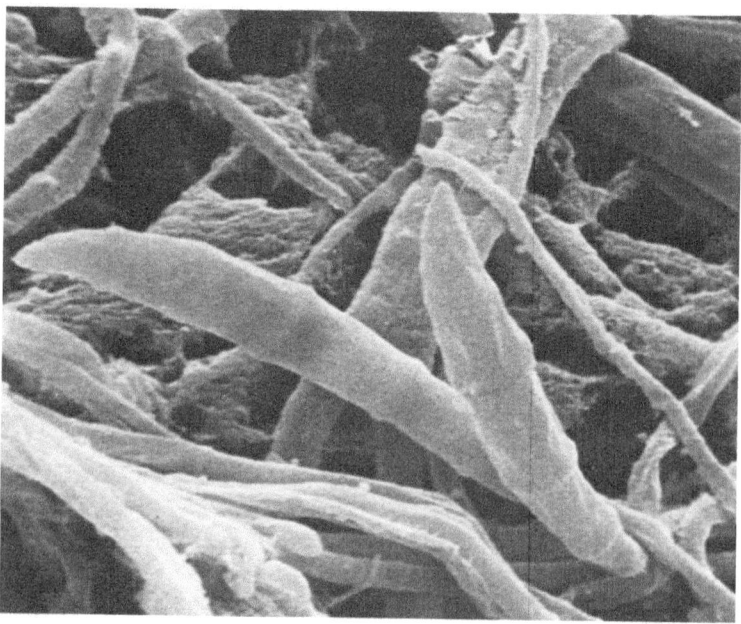

Fig. 2.8. Scanning electron micrograph of mature macroconidia of *Fusarium culmorum* in which the septa have been formed. × 2200.

has to be established that this mechanism described from a submerged culture is identical with the mechanism on solid media. With regard to the last point all the available indications are that there are no differences associated with different growth conditions.

A very recent paper has been published by Schneider & Seaman (1982*b*) devoted to the ontogeny of conidia in *F. sulphureum*. The bulk of this paper deals with an ultrastructural study of macroconidia produced on a solid medium and the basic conclusion reached by the authors is that the conidia are produced through an enteroblastic mechanism. They suggest that an extension of the inner wall layer at the tip of the phialide emerges through the degrading outer wall layer in a manner compatible with the concept of an enteroblastic system (Minter, Kirk & Sutton, 1982). The inevitable consequences of the mechanism illustrated by Schneider & Seaman (1982*b*) are that the expanding conidium emerges through its enveloping wall layer at an early stage, while it is still small in size, and therefore the collarette is visible early in the ontogeny. Also the mechanism of emergence of the expanding conidium, figured by these authors, does not lead to the formation of a cap of material over the conidium tip. Although there seems no doubt over the basic interpretation of the ontogeny in *F. sulphureum* the authors have used an extreme fixation procedure and any detailed interpretation, particularly of wall structures, could be questioned. However, there remains the conflict of published data from several species since all the information cannot be reconciled into a single ontogenetic sequence. Several possibilities exist, first since *F. sulphureum* is the anamorph of a species of *Gibberella*, *F. decemcellulare* the anamorph of a species of *Calonectria*, and *F. culmorum* has no known teleomorph, the differences may lie in the taxonomic dispositions of the organisms. Second since different growth regimes have been used we may be dealing with environmentally induced differences in ontogeny, a somewhat unlikely possibility. Third one or more of the sets of data have been wrongly interpreted, which is always likely; yet even if this were the case the ultrastructural images would still require reinterpretation. The whole problem is complicated by the fact that Schneider & Seaman (1982*b*) claim to have examined *F. culmorum* and *F. decemcellulare* and to have found them similar to *F. sulphureum* although they produce no evidence.

We are thus left in the position that Schneider & Seaman clearly believe that the work on *F. culmorum* and *F. decemcellulare* is in some way incorrect and yet my own observations, using scanning electron

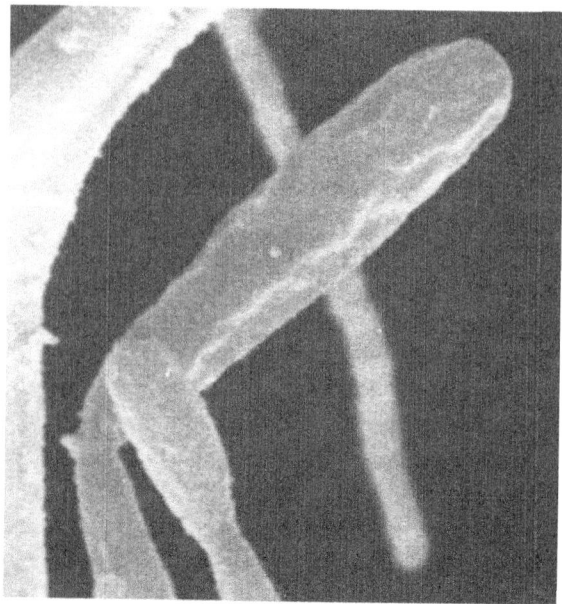

Fig. 2.9. Scanning electron micrograph of young macroconidia of *Fusarium culmorum*. Note that no collarette is visible at the base of the spores. × 4400.

microscopy, on *F. culmorum* grown on solid medium fail to show the collarette around the base of the conidium at an early stage of development (Fig. 2.9), which would be necessary to support the fully enteroblastic ontogeny. Regrettably we are therefore left with no clear resolution of the dilemma at this time and have to await further experimental information.

The development of chlamydospores

Many species of *Fusarium* produce conidial or hyphal chlamydospores, whose importance for the organism's survival in the soil has been investigated (Sitton & Cook, 1981). Superficial observation of mature macroconidia of many species of *Fusarium* shows that one of the first observable changes on ageing, particularly in adverse conditions, is the loss of cytoplasmic material from the apical and basal cells (Schneider & Seaman, 1974a). Following this, changes occur in the wall structure leading to an overall thickening and the individual conidial cells become rounded. The wall thickening clearly results from the addition of a new inner wall layer, irrespective of the interpretation placed on the number of layers in the original conidial wall (Campbell & Griffiths, 1974; Stevenson & Becker, 1972). The swelling and rounding-up of the developing conidial chlamydospores can lead to the rupture of the outer wall derived from the conidial wall (Campbell & Griffiths, 1974) or to the production of a fibrillar outer layer, also presumably arising from conidial wall destruction (Stevenson & Becker, 1972). As with many resistant spores the chlamydospores become more difficult to fix as they mature, but despite this some internal changes can be observed. Schneider and Seaman (1974*b*) have reported a 'cellular body' formed during differentiation of conidial chlamydospores in *F. sulphureum* and *F. poae*. The structure they refer to is derived from endoplasmic reticulum membrane which becomes dilated and contains granular material and microbodies. It is of course impossible to ascribe functions accurately to such observed structures and since similar membrane forms have previously been illustrated in normal mature germinating conidia (Marchant, 1966*a*) it is doubtful whether their function is specific to chlamydospore formation.

Conidial chlamydospores, in common with macroconidia, contain lipid; in some instances when the chlamydospores are produced from 'high reserve' macroconidia they may contain large amounts of lipid in a single storage structure. The presence of such large amounts of lipid does not, however, seem to improve the survival potential of chlamydospores of *F. solani* (van Eck, 1978*a*). Since chlamydospores seem destined to survive for long periods in soil it is clearly of interest to observe what changes, if any, take place over a period of time. In a number of species the interaction of soil microorganisms with the spores leads to progressive degradation of the outer wall layers and lysis of the cytoplasmic content, although without apparent penetration of the spores by microorganisms (van Eck, 1976).

Cell wall structure

The majority of fungal cell walls have chitin, organised into crystalline microfibrillar structures, as a component of variable proportion. The matrix material in which these fibrils are embedded is normally composed largely of amorphous polysaccharide material, usually predominantly glucans of various types. Chemical analysis of the walls of various fusaria reveals the presence of glucosamine, glucose, mannose and galactose (Schneider *et al.* 1977; van Eck, 1978*b*). The glucosamine is derived, at least in part, from chitin; a fact which can be demonstrated through the use of X-ray diffraction analysis (Schneider *et al.*, 1977). This technique can also be used to show that there is an increase in the degree of crystallinity of the chitin in walls of conidia of *F. sulphureum* during their maturation after release (Schneider & Seaman, 1982*a*). In addition to the hexosamine and hexose content, Marchant (1966*b*) reported the presence of pentose material in the walls of macroconidia of *F. culmorum* which was ascribed to the mucilage layer. Glucuronic acid in substantial quantity has also been reported from the walls of spores of *F. sulphureum* (Schneider *et al.*, 1977). Spore walls also regularly contain between 20–30% peptide material (Laborda *et al.*, 1974; Schneider *et al.*, 1977), which on hydrolysis yields a fairly complete spectrum of amino acids (Marchant, 1966*b*). The only other major component is lipid which contributes less than 5% of the wall material (Laborda *et al.*, 1974; Schneider *et al.*, 1977). The nature of the mucilage layer at the surface of the macroconidia of two species of *Fusarium* has been examined using lectin-binding techniques by Kleinschuster & Baker (1974). Although not providing direct information on the actual identity of the surface components this work has indicated that there are detectable differences between species.

Further indirect evidence on the structure of walls can be obtained through looking at the lysis of walls by enzyme preparations to yield spheroplasts and protoplasts. Through a knowledge of the enzyme components in the preparation used and its effectiveness in producing protoplasts the direct chemical analysis information can be confirmed and expanded (García Acha, López-Belmonte & Villanueva, 1966).

However complete our knowledge of the chemical composition of the spore walls this information does not tell us how the components are structurally organised in the wall. Direct observation of isolated wall material, using shadowing techniques in the electron microscope, fails to show the presence of microfibrils and these can only be revealed after quite drastic chemical treatment (Schneider & Wardrop, 1979). Using this methodology the conidial wall can be shown to consist of a single layer of randomly oriented chitin microfibrils embedded in an amorphous matrix, with the microfibrillar layer of the outer conidial wall continued through the septum (Schneider & Wardrop, 1979). The additional wall layer produced during the formation of chlamydospores also contains microfibrillar material (Griffiths, 1973) and since it can be shown that chlamydospore walls contain a greater percentage of glucosamine and peptide material and less hexose polymer material than conidia (Schneider *et al.* 1977) it seems reasonable to propose that the chlamydospore wall thickening consists largely of chitin rather than matrix material.

The physiology of sporulation

Once again, although there have been many papers relating to the biochemical activities of fusaria, the process of sporulation has been largely neglected. The biochemical processes associated with spore production are extremely difficult to study when the organism is growing on solid medium, and much of the literature deals with this type of system. Some interest in sporulation in submerged liquid cultures began to occur about twenty years ago (Cappellini & Peterson, 1965). The most easily manipulated system for studying sporulation in sub-merged culture is the chemostat where the organism can be maintained in a steady, although variable, metabolic state related to a variety of variables such as growth rate, substrate composition, and type of substrate limitation. One of the best examples of exploitation of this growth system is shown by the work of Smith and his colleagues on *Aspergillus niger* (Smith & Anderson, 1973). These workers have also developed a so-called microcycle conidiation system in which the development of a heterogeneous mycelial population is circumvented and conidia can be induced to germinate and produce conidia directly. As we shall see there is evidence that a similar system could be used with fusaria.

Specific studies of spore production in fusaria have often been preoccupied with the factors and conditions which lead to the formation of spores and the relative numbers produced. Such an approach leads to a wealth of literature which is difficult to interpret and in the final analysis provides us with little information on the basic metabolic and differentiation patterns underlying sporulation. It is clear from a wide range of studies of fungal sporulation that no single factor acts as a universal trigger for sporulation. For example in the paper by Hsieh, Snyder & Smith (1979) the influence of carbon sources, amino acids and water potential on growth and sporulation are examined. Many factors have been suggested as triggers for sporulation in fungi and these include low growth rate and nitrogen depletion (Righelato *et al.* 1968); nitrogen source (Anderson & Smith, 1971*a*); temperature and oxygen concentration (Anderson & Smith, 1971*b*); pH (Zendler & Margalith, 1972); carbon source (Audhya & Russell, 1974); glucose concentration (Morton, 1961) and carbon dioxide concentration (Graafmans, 1974). Since most of these cited studies have employed surface cultures, the interactions of the various factors are difficult to define: for example declining growth rate (which is often taken to lead to spore production) cannot be separated clearly from the exhaustion of carbon or nitrogen sources, and similarly localised changes in gas concentration or pH can occur in the non-homogeneous cultures on solid media.

At least some of the interactions can be separated when a chemostat system is used, since growth rate can be controlled independently of nutrient supply, and dissolved gas concentrations and pH can also be controlled more rigorously. The work of Larmour & Marchant (1977a, b) has shown that cultures of *F. culmorum* can successfully be grown and sporulated in continuous culture. Larmour and Marchant grew *F. culmorum* from a macroconidial inoculum in a glucose mineral salts medium in a chemostat at dilution rates up to 0.1 h^{-1} under various substrate concentrations and limitations. If the culture was left without pH control it achieved a steady pH of approximately 3.5 and the organism remained entirely filamentous and unpigmented and produced no spores. The low pH was maintained as a result largely of selective ion uptake and partly through the production of acidic metabolites (Katouli & Marchant, 1981). The non-sporulating condition was maintained indefinitely under the conditions of low pH. The culture became pigmented and produced abundant macroconidia within a few hours of the pH being raised and controlled at a value of 6.5. The critical threshold for conidial production was a pH above 3.8 for 6 hours or more. The rate of conidial development was independent of both the pH and dilution rate, but the rate of conidial production was greatest at the highest growth levels. In phosphate- and nitrogen-limited cultures increasing glucose concentration progressively suppressed conidiation. The major physiological change which was detected in the cultures was an approximately nine-fold increase in the rate of carbon dioxide fixation in conidiating cultures, which could be explained partly by increases in activity of phosphoenolpyruvate carboxykinase (EC 4.1.1.32) and pyruvate carboxylase (EC 6.4.1.1). Thus, in the *F. culmorum* system, type of substrate limitation and growth rate do not seem to be the major triggers for sporulation and pH can be used as a seemingly indirect trigger for the process via anaplerotic carbon fixation pathways.

Lipid composition

An obvious constituent of both the mycelia and spores of fusaria is lipid, often in the form of discrete organelles. Some efforts

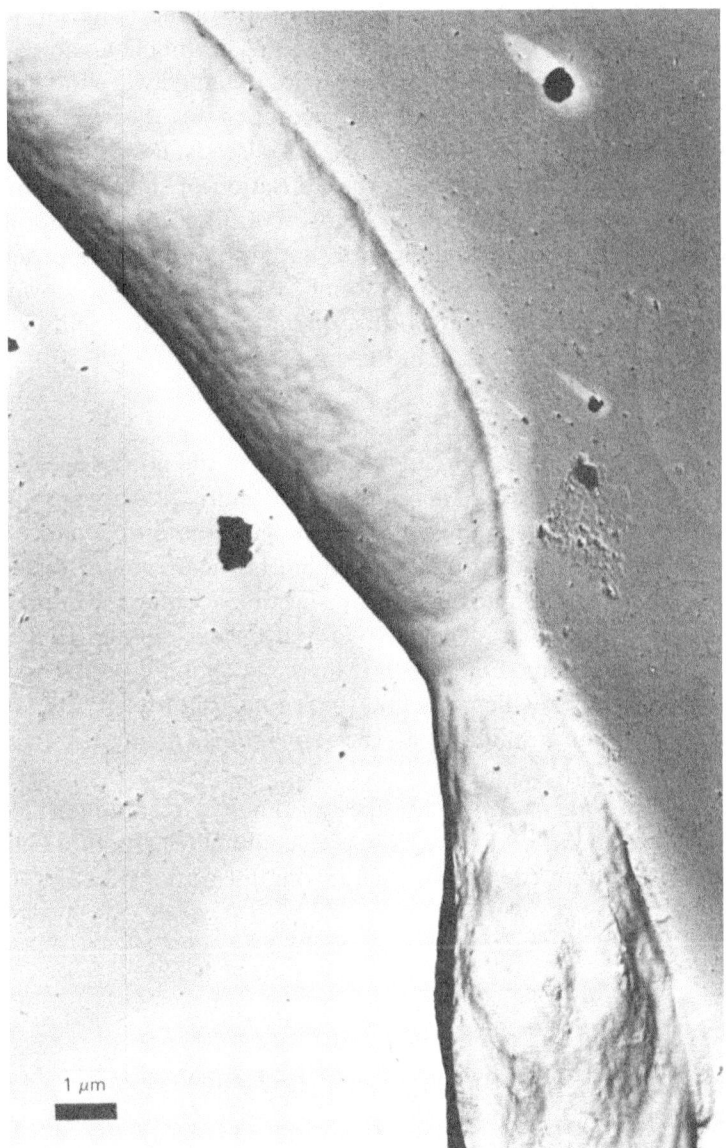

1 μm

Fig. 2.10. Replica of anomalous spore germination in *Fusarium culmorum* in which a macroconidium is germinating to produce a new spore directly. Note that both the germinating spore and the newly formed spore appear to have a covering of mucilage.

have been made to examine the composition of this material (Starratt & Madhosingh, 1967; Rambo & Bean, 1969; Nombela-Cano & Peberdy, 1971). In each case the major fatty acids detected were the unsaturated 18:1 and 18:2 acids with smaller quantities of other eighteen carbon, sixteen and fourteen carbon acids. As already indicated Mumford & Pappelis (1978) have followed the utilisation of this lipid during spore germination and we might therefore consider that the formation of elevated quantities of lipid is an essential feature of the production of spores in fusaria. Surprisingly Rambo and Bean have shown that for *F. oxysporum* and *F. roseum* the amount of the fatty acids was always higher in the mycelia than in the spores.

Anomalous spore production

Reference has already been made to the microcycle sporulation system developed for *Aspergillus niger* (Smith & Anderson, 1973). In most instances macroconidia of fusaria germinate to produce a mycelium which only later produces conidia; however, under certain conditions an unusual sequence of events occurs in which the germ tube fails to develop into a branched hypha and instead immediately produces another conidium (Fig. 2.10) (Marchant, unpublished results). The conditions under which this takes place have not been fully examined, but it bears some similarity to the *Aspergillus* situation.

Acknowledgement. I would like to thank Dr. Stephen F. Lowry, Experimental Officer in the New University Ultrastructure Unit, for his extensive help in the preparation of the illustrations for this chapter.

References

Akai, S., Fukutomi, M. & Kobayashi, N. (1969). Fine structure of conidia of *Fusarium oxysporum* f. *cucumerinum. Annals of the Phytopathological Society of Japan*, **35**, 351–53.

Anderson, J. G. & Smith, J. E. (1971a). Synchronous initiation and maturation of *Aspergillus niger* conidiophores in culture. *Transactions of the British Mycological Society*, **56**, 9–29.

Anderson, J. G. & Smith, J. E. (1971b). The production of conidiophores and conidia by newly germinated conidia of *Aspergillus niger* (microcyle conidiation). *Journal of General Microbiology*, **69**, 185–97.

Audhya, T. K. & Russell, D. W. (1974). Production of enniatins by *Fusarium sambucinum*: selection of high-yield conditions from liquid surface cultures. *Journal of General Microbiology*, **82**, 181–90.

Barran, L. R., Schneider, E. F. & Seaman, W. L. (1977) Requirements for the rapid conversion of macroconidia of *Fusarium sulphureum* to chlamydospores. *Canadian Journal of Microbiology,* **23,** 148–51.

Booth, C. (1971). *The Genus* Fusarium. Kew: Commonwealth Mycological Institute.

Campbell, W. P. & Griffiths, D. A. (1974). Development of endoconidial chlamydospores in *Fusarium culmorum. Transactions of the British Mycological Society,* **63,** 221–8.

Cappellini, R. A. & Peterson, J. L. (1965). Macroconidium formation in submerged cultures by a non-sporulating strain of *Giberella zeae. Mycologia,* **57,** 962–6.

Carroll, G. C. (1969). A study of the fine structure of ascosporogenesis in *Saccobolus kerverni. Archiv für Mikrobiologie,* **66,** 321–39.

Cole, G. T. & Samson, R. A. (1979). *Patterns of Development in Conidial Fungi.* London: Pitman.

Ford, E. J., Gold, W. C. & Snyder, W. C. (1970*a*). Induction of chlamydospore formation in *Fusarium solani* by soil bacteria. *Phytopathology,* **60,** 479–84.

Ford, E. J., Gold, W. C. & Snyder, W. C. (1970*b*). Interaction of carbon nutrition and soil substances in chlamydospore formation by *Fusarium. Phytopathology,* **60,** 1732–7.

García Acha, I., Aguirre, M. J. R., Uruburu, F. & Villanueva, J. R. (1966). The fine structure of the *Fusarium culmorum* conidium. *Transactions of the British Mycological Society,* **49,** 695–702.

García Acha, I., López-Belmonte, F. & Villanueva, J. R. (1966). Preparation of protoplast-like structures from conidia of *Fusarium culmorum. Antonie van Leeuwenhoek,* **32,** 299–311.

Goos, R. D. & Summers, D. F. (1964). Use of fluorescent antibody techniques in observations on the morphogenesis of fungi. *Mycologia,* **56,** 701–7.

Graafmans, W. D. J. (1974). Metabolism in *Penicillium isariiforme* on exposure to light with special reference to citric acid synthesis. *Journal of General Microbiology,* **82,** 247–52.

Griffin, G. J. (1976). Roles of low pH, carbon and inorganic nitrogen source use in chlamydospore formation by *Fusarium solani. Canadian Journal of Microbiology,* **22,** 1381–9.

Griffiths, D. A. (1973). Fine structure of the chlamydospore wall in *Fusarium oxysporum. Transactions of the British Mycological Society,* **61,** 1–6.

Hsieh, W. H., Snyder, W. C. & Smith, S. N. (1979) Influence of carbon sources, amino acids, and water potential on growth and sporulation of *Fusarium moniliforme. Phytopathology,* **69,** 602–4.

Katouli, M. & Marchant, R. (1981). Effect of phytotoxic metabolites of *Fusarium culmorum* on growth and physiology of barley plants. *Plant and Soil,* **60,** 377–84.

Kendrick, B. (ed.) (1971). *Taxonomy of Fungi Imperfecti.* Toronto: University of Toronto Press.

Kleinschuster, S. J. & Baker, R. (1974). Lectin-detectable differences in carbohydrate-containing surface moieties of macroconidia of *Fusarium roseum* 'Avenaceum' and *Fusarium solani. Phytopathology,* **64,** 394–9.

Laborda, F., García Acha, I., Uruburu, F. & Villanueva, J. R. (1974). Structure of conidial walls of *Fusarium culmorum. Transactions of the British Mycological Society,* **62,** 557–66.

Larmour, R. & Marchant, R. (1977*a*). The induction of conidiation in *Fusarium culmorum* grown in continuous culture. *Journal of General Microbiology,* **99,** 49–58.

Larmour, R. & Marchant, R. (1977*b*) Carbon dioxide fixation and conidiation in

Fusarium culmorum grown in continuous culture. *Journal of General Microbiology*, **99**, 59–68.

Marchant, R. (1966a). Fine structure and spore germination in *Fusarium culmorum*. *Annals of Botany*, **30**, 441–5.

Marchant, R. (1966b). Wall structure and spore germination in *Fusarium culmorum*. *Annals of Botany*, **30**, 821–30.

Marchant, R. (1975). An ultrastructural study of 'phialospore' formation in *Fusarium culmorum* grown in continuous culture. *Canadian Journal of Botany*, **53**, 1978–87.

Marchant, R. (1979). Wall growth during spore differentiation and germination. In *Fungal Walls and Hyphal Growth*, ed. J. H. Burnett & A. P. J. Trinci, pp. 115–48. Cambridge University Press.

Mason, E. W. (1933). Annotated account of fungi received at the Imperial Bureau of Mycology. List II. *Mycological Papers*, **3**, 1–67.

Messiaen, C.-M. & Cassini, R. (1968). Recherches sur les fusarioses IV. La systématique des *Fusarium*, *Annals Epiphyties*, **19**, 387–454.

Minter, D. W., Kirk, P. M. & Sutton, B. C. (1982). Holoblastic phialides. *Transactions of the British Mycological Society*, **79**, 75–93.

Morton, A. G. (1961). The induction of sporulation in mould fungi. *Proceedings of the Royal Society, B*, **153**, 548–69.

Mumford, P. M. & Pappelis, A. J. (1978). Dry mass of *Fusarium roseum* spores before and after germination. *Mycopathologia*, **64**, 63–4.

Nombela-Cano, C. & Peberdy, J. F. (1971). The lipid composition of *Fusarium culmorum* mycelium. *Transactions of the British Mycological Society*, **57**, 342–4.

Rambo, G. W. & Bean, G. A. (1969). Fatty acids of the mycelia and conidia of *Fusarium oxysporum* and *Fusarium roseum*. *Canadian Journal of Microbiology*, **15**, 967–8.

Reichle, R. E. & Alexander, J. V. (1965). Multiperforate septations, Woronin bodies, and septal plugs in *Fusarium*. *Journal of Cell Biology*, **24**, 489–96.

Righelato, R. C., Trinci, A. P. J., Pirt, S. J. & Peat, A. (1968). The influence of maintenance energy and growth rate on the metabolic activity, morphology and conidiation of *Penicillium chrysogenum*. *Journal of General Microbiology*, **50**, 399–412.

Schneider, E. F., Barran, L. R., Wood, P. J. & Siddiqui, I. R. (1977). Cell wall of *Fusarium sulphureum*. II. Chemical composition of the conidial and chlamydospore walls. *Canadian Journal of Microbiology*, **23**, 763–9.

Schneider, E. F. & Seaman, W. L. (1974a) Development of conidial chlamydospores of *Fusarium sulphureum* in distilled water. *Canadian Journal of Microbiology*, **20**, 247–54.

Schneider, E. F. & Seaman, W. L. (1974b). Development of a cellular body during differentiation of conidial chlamydospores in *Fusarium*. *Canadian Journal of Microbiology*, **20**, 1205–8.

Schneider, E. F. & Seaman, W. L. (1982a). Structure of chitin in the cell walls of newly formed and mature conidia of *Fusarium sulphureum*. *Canadian Journal of Microbiology*, **28**, 531–5.

Schneider, E. F. & Seaman, W. L. (1982b). Ontogeny of conidia in *Fusarium sulphureum*. *Transactions of the British Mycological Society*, **79**, 283–90.

Schneider, E. F. & Wardrop, A. B. (1979). Ultrastructural studies on the cell walls in *Fusarium sulphureum*. *Canadian Journal of Microbiology*, **25**, 75–85.

Seemüller, E. (1968). *Arbeiten aus der biologischen Bundesanstalt für Land- u. Forstwirtschaft*, **127**, 1–93.

Sitton, J. W. & Cook, R. J. (1981). Comparative morphology and survival of chlamydospores of *Fusarium roseum* 'Culmorum' and 'Graminearum'. *Phytopathology*, **71**, 85–90.

Smith, J. E. & Anderson, J. G. (1973). Differentiation in the Aspergilli. In *Microbial Differentiation*, ed. J. M. Ashworth & J. E. Smith, pp. 295–337. Twenty-third Symposium of the Society for General Microbiology. Cambridge University Press.

Šrobárová, A. & Šrobár, S. (1979). Ultraštruktúra konídií *Fusarium culmorum* (W.G. Sm.) Sacc. *Biologia (Bratislava)*, **34**, 587–91.

Starratt, A. N. & Madhosingh, C. (1967). Sterol and fatty acid components of mycelium of *Fusarium oxysporum*. *Canadian Journal of Microbiology*, **13**, 1351–5.

Stevenson, I. L. & Becker, S. A. W. E. (1972). The fine structure and development of chlamydospores of *Fusarium oxysporum*. *Canadian Journal of Microbiology*, **18**, 997–1002.

Subramanian, C. V. (1971). The phialide. In *Taxonomy of Fungi Imperfecti*, ed. B. Kendrick, pp. 92–119. University of Toronto Press.

Tewari, J. P. & Skoropad, W. P. (1975). Fine structure of the macroconidia of *Fusarium solani*. *Canadian Journal of Botany*, **53**, 2134–46.

van der Valk, P., Marchant, R. & Wessels, J. G. H. (1977). Ultrastructural localization of polysaccharides in the wall and septum of the basidiomycete *Schizophyllum commune*. *Experimental Mycology*, **1**, 69–82.

Van Eck, W. H. (1976) Ultrastructure of forming and dormant chlamydospores of *Fusarium solani* in soil. *Canadian Journal of Microbiology*, **22**, 1634–42.

Van Eck, W. H. (1978a). Lipid body content and persistence of chlamydospores of *Fusarium solani* in soil. *Canadian Journal of Microbiology*, **24**, 65–9.

Van Eck, W. H. (1978b). Chemistry of cell walls of *Fusarium solani* and the resistance of species to microbial lysis. *Soil Biology and Biochemistry*, **10**, 155–7.

Van Eck, W. H. & Schippers, B. (1976). Ultrastructure of developing chlamydospores of *Fusarium solani* f. *cucurbitae in vitro*. *Soil Biology and Biochemistry*, **8**, 1–6.

Wergin, W. P. (1973). Development of Woronin bodies from microbodies in *Fusarium oxysporum* f. sp. *lycopersici*. *Protoplasma*, **76**, 249–60.

Zendler, G. & Margalith, P. (1972). Synchronized sporulation in *Penicillium digitatum* (Sacc.). *Canadian Journal of Microbiology*, **18**, 1685–90.

3
Aspects of *Fusarium* genetics

J.H.BURNETT

University of Edinburgh, Old College, South Bridge, Edinburgh EH8 9YL, U.K.

Introduction

It is probably an historical accident that *Fusarium* species have not been much used by geneticists but is not difficult to find reasons for continuing to leave them well alone! It is possible to study the genetics of organisms whose taxonomic status is unknown or uncertain provided that the organisms can readily be recognised. It is much more difficult when their range of variation overlaps with that of other very similar organisms of equally uncertain status, or when they co-exist with morphologically identical organisms from which, to varying degrees, they are isolated genetically. Moreover, if sexual reproduction is totally lacking or apparently so, while some of the anamorphic forms turn out to be associated with at least five genetically distinct teleomorphs (some differing in chromosome number), the problems become greater. It is the genetics of a complex that is being studied rather than that of a species.

On the other hand, to apply genetics to elucidate the causes of variation in a notoriously variable group, to analyse the differences between the saprophytic and the pathogenic condition and to investigate the basis of virulence and host specificity cannot but be challenging to the mycogeneticist.

In order to bring some coherence to the present treatment and to avoid mere cataloguing, I shall concentrate on two species, as defined by Snyder and Hansen (1940), *Fusarium oxysporum* and *F. solani*.[1] I realise that this may not be taxonomically satisfactory but it does have the double advantage of dealing with entities recognised and employed

[1] Nomenclature used in this chapter follows Booth (1971).

by pathologists and of demonstrating the range of problems that *Fusarium* genetics presents. Puhalla (1981) has recently reviewed formal genetical aspects of fusaria.

These two species are amongst the most widespread and predominant in natural and cultivated soils. Both species include vigorous saprophytes which may colonise senescent or damaged plant tissues as well as virulent pathogens involved in root and crown rots and, in the case of *F. oxysporum*, serious vascular wilts. A sexual phase (teleomorph) of *F. oxysporum*, is not known but some strains of *F. solani* do produce perithecia in nature, especially on aerial plant parts in the tropics. However, it has been shown that failure to produce perithecia over wide areas can also result from the absence of a complementary strain. Perithecia can be produced readily in the laboratory if appropriate strains are paired.

The two species are defined on morphological grounds. Saprophytic strains are so described but the pathogenic strains are further subdivided, principally on the basis of their selective pathogenicities. Thus some 74 formae speciales are now recognised for *F. oxysporum* (Armstrong & Armstrong, 1980) and ten for *F. solani* (Matuo & Snyder, 1973) while, in addition, in some formae speciales, two or more races are recognised by their reaction with differential varieties of the host plants. In all, something of the order of 110–120 forms of *F. oxysporum* and 15–20 of *F. solani* are recognised by the joint application of morphological and biological criteria.

Morphological variation

Both species show considerable morphological variation. Some of it is environmentally induced but a large amount also arises from genetic changes, especially during prolonged culture. Booth (1975) has commented that: 'It can be categorically stated that, if fusaria are isolated from fructifications on plant tissue or grown out of plant tissue on neutral media without antibiotics, and single spore cultures taken as soon as spores appear, the range of single spore isolates will show no more variation than would be expected of any other microfungi.'

Phenotypic plasticity

In culture, the concentration and composition of growth media and their pH values have long been known to affect features such as growth rate; the shape, size and abundance of spores, sclerotia or sexual reproductive structures; the numbers of septa in macroconidia; and pigmentation, both internal and that released into the medium (e.g. Sideris, 1925; Brown & Horne, 1926; Horne & Mitter, 1927; Snyder & Hansen, 1941*a*; Carlile, 1956). Light and temperature also affect the production and dimensions of spores and protoperithecia as well as pigmentation (e.g. Smith & Swingle, 1904; Harter, 1939; Snyder & Hansen, 1941*b*). In order to detect genuine genetic differences it is, therefore, crucial to control cultural and environmental conditions as fully as possible and to keep them constant.

The adaptive significance of the great potential phenotypic plasticity of fusaria and, indeed, how far it is manifest in the natural environment is not clear. Indeed, until a more detailed knowledge of their ecology is available, it is likely to remain obscure. One or two features are obviously adaptive. For example, the production of sporodochia and perithecia in response to light, in those species exhibiting either splash dispersal of spores at the soil surface or aerial dispersal of spores, is an obvious adaptive response for a soil-based species such as *F. solani*. What is not made clear by such an example is why the fusaria appear to exhibit more phenotypic plasticity than is shown by a great many other fungi growing in similar habitats.

Genetic changes

A well-defined range of morphological variation has been recognised for many years in some isolates of *F. oxysporum* and to a lesser extent in *F. solani*. A classification of the appearance of these variants is given in Table 3.1. Forms intermediate between these are also known. All such variants can be isolated from diseased material or soil. They are most usually detected as sectors in cultures, as an apparently gradual transformation over successive mass cultures, or abruptly – as rare events – when single, uninucleate microconidia are isolated and grown on.

The phenotypes do not arise in culture at random (Hwang, 1948). *co* forms can give rise to *sp*, *ro*, *slp* or *sh* forms; the changes may be sequential but are not necessarily so. Reversion in culture is unusual and *co* and *sh* are generally exceedingly stable, but revertants have been isolated from host inoculations, e.g. *sp* from *slp* in *F. oxysporum* f. sp. *cubense* (Follin & Laville, 1966). The frequency of production of novel phenotypes is affected both by the medium and by the physiological age of the propagule used for subculturing. On the whole the higher the C and N content of the medium (e.g. so-called richer media such as Richard's or Potato dextrose agar), the higher the frequency of sector-

Table 3.1. *Five common morphological variant types found in fusaria. Intermediates and lesser variants also occur*

			Morphological variants			
Characteristic feature	Sporodochial *sp*	Cottony *co*	Ropy *ro*	Slimy pionnotal *slp*	Shorn *sh*	
Aerial mycelium	Abundant, floccose	Very abundant, fine, cottony	Very abundant, ropy, often in raised tufts	Scattered, apressed, slimy	Virtually none	
Pigment	Pink or purple	Pale pigments (white)	Highly pigmented	Highly pigmented	Highly pigmented	
Sporodochia	Erumpent or submerged	Absent	Absent	Absent	Absent	
Pionnotes	Present	Few to absent	Absent	Abundant	Absent to few	
Commonest change if not stable	To *co* or *slp*	To *sp*	Highly unstable to *slp* or *sh*	To *sh*	—	

Based on Waite & Stover, 1960 and Follin & Laville, 1966.

ing (Brown, 1926; Miller, 1946). Follin & Laville (1966) have shown that, even starting with a single uninucleate microconidium, all the morphological variants can be produced but that the most stable colonies are those propagated either by hyphal tips transferred at 4-day intervals, or by microconidia at about 20-day intervals. Buxton (1954), having isolated the different morphological variants from sectors of *F. oxysporum* f. sp. *gladioli*, was also able to maintain them for several hundred subcultures (with only a single detected variant in their progeny) by using microconidial transfers. By contrast, Follin and Laville employed 20-day transfers of macroconidia, chlamydospores and 40–60 μm fragments of old hyphae. Here variants arose at the second or third transfer of *sp* cultures, either to *slp* or to *ro* phenotypes, in the case of old hyphae and chlamydospores, but they arose somewhat less rapidly from macroconidial transfers. These authors note that there is a positive correlation between the age of the propagule and the probability of a variant arising, i.e. the older it is the more likely a variant becomes. The 'age' effect was further supported by noting that although apical tip and microconidial transfers were stable over the three-month experimental period, colonies derived from them did show typical signs of senescence. Their hyphae became more vacuolated, growth was slower and, some 50 days from the start, chlamydospores ceased to form. Mass transfers from the colonies taken between 25 and 90 days tended to develop *sp* features, and from 90–120 days *rp* phenotypes occurred as well. Thus the internal as well as the external environment either induces, or merely selects, these changes. Experimental data on these matters are sparse and inadequate. The changes induced could be either cytoplasmic or chromosomal but, at present, the latter site is favoured. However, it should be pointed out that similar variants have arisen in *F. oxysporum* when treated with acriflavine, an agent regularly implicated in the induction of extra-chromosomal variants (Singh, 1973). Spontaneous variants *A*(ring) and *S*(sector) in self-fertile *Nectria haematococca* also have a cytoplasmic base. Both *A* and *S* differ from wild type mycelia in having fewer aerial hyphae, a reduced growth rate and the release of dark pigment: they therefore resemble the *slp* phenotype in several respects. The agents responsible for these phenotypes are infective and non-nuclear (Bouvier & Laville, 1970) but their expression and suppression can be affected by nuclear genes (Bareyre & Laillier-Rousseau, 1972; Daboussi-Bareyre, 1976, 1977, 1980; Daboussi-Bareyre, Laillier-Rousseau & Parisot, 1979).

Table 3.2. *Ratios of homokaryotic types F and A recovered from heterokaryon derived from 1 : 1 mixed microconidial inoculum on water agar after transference to four media differing in C : N ratio and growth to colonies 1.5 cm in diameter*

Glucose: Peptone ratio (N constant at 19 gm/1 peptone)	Microconidial isolates		Ratio
	Type F[1]	Type A[2]	
Water agar (0 : 0)			0.5
0 : 19	343	479	0.71
2.4 : 19	158	499	0.33
4.8 : 19	21	1.82	0.11
7.2 : 19	132	464	0.28

After Buxton, 1954.
[1] Adpressed, somewhat ropy, no sporodochia, pionnotes in light (*slp*).
[2] Abundant, cottony mycelium, abundant sporodochia, no pionnotes (*co*).

Nevertheless, in favour of a chromosomal basis, mutants resembling the different morphological phenotypes can be produced by uv-irradiation (Buxton, 1956), or with nitrous acid, ethane methanesulphonate and nitrosoguanidine (Sanchez, Leary & Endo, 1975) – all of which affect chromosomal loci. In many strains, however, and under appropriate conditions, such as low temperature, low-nutrient media or soil-tube culture (Miller, 1946), both normal and variant phenotypes are completely stable. Thus the red sporodochial, purple cottony and white apressed homokaryons of *F. moniliforme* are quite stable in culture (Ming, Lin & Yu, 1966). Moreover, using *F. oxysporum* f. sp. *gladioli*, Buxton (1954) has not merely been able to make heterokaryons between different, spontaneous morphological variants but also to recover them intact. Indeed, by varying the relative concentrations of carbon or nitrogen constituents in the medium, he was able to alter the nuclear ratios of heterokaryons and, hence, to select for a particular genotype (Tables 3.2 and 3.3). Experimental verification of his data is necessary since his work is at present unique. Finally, in the sexually reproducing *F. solani*, mutation at the mutable *c* locus (to be described in the next section) results in the loss of both protoperithecia and sporodochia; in a wholly anamorphic form this would resemble the *co* phenotype.

Their importance for classification alone makes it evident that a more

Table 3.3. *Ratios of homokaryotic types F and A recovered from a heterokaryon derived as in Table 2 after isolating 21 hyphal tips and transferring to either a high or a low C:N medium*

	Ratio F:A*										
High C:N (2.0%:0.01%)	0.36	0.37	0.41	0.44	0.50	0.53	0.57	0.62	0.70		
Low C:N (0.2%:0.1%)	0.89	1.01	1.01	1.26	1.33	1.37	1.42	1.54	1.56	2.1	3.4	3.67

After Buxton, 1954.
* A 't' test gave $t_{(19)} = 3.96$, significant at 1% level.

precise genetic analysis of such morphological variation is essential. Incidentally, with the increasing use of some fusaria for industrial fermentation or protein production, an understanding of strain stability, mutability of hyphal form, and sporulation could be of great economic importance.

Recombination systems

As stated earlier some fusaria reproduce, or are potentially capable of reproducing, sexually; others are apparently entirely asexual. In general, fusaria seem to be a group in which there is a tendency for sexuality to be lost but it is not clear whether such loss is associated with a compensatory development of other recombinant systems.

Sexuality

Self-fertile forms, not surprisingly, produce perithecia most commonly, e.g. *F. graminearum* Schwabe (= *Gibberella zeae* (Schw., Petch). However, both in E. Australia and S.W. France two extremely similar populations co-exist in the same soils although one of them usually predominates. One form rarely produces perithecia in nature, or in normal culture, whereas the other produces them in all conditions. Isolates resembling the former arise and can be selected from the French perithecium-forming population (Francis & Burgess, 1977; Messiaen & Cassini, 1981), but the basis of the change is not known.

Some species include both self-fertile and self-sterile forms which appear to be crossable, e.g. *F. decemcellulare* Brick (= *Calonectria rigidiuscula* (Berk. & Br.) Sacc.), a pathogen of cocoa (Reichle & Snyder, 1964). In this and other cases the self-fertile forms are said to be largely saprophytes and the self-sterile ones to be pathogens e.g. *F. solani* (Snyder & Hansen, 1941*a*).

There is also considerable variation in perithecium production amongst self-sterile species. *F. moniliforme* (= *Gibberella fujikuroi* (Sawada) Wr.) is said to produce perithecia commonly in subtropical areas but rarely in Europe (Messiaen & Cassini, 1981). Similarly the perithecia of the widespread *F. solani* f. sp. *pisi* have been found only infrequently in nature in Japan, on wet lesions on mulberry (Sakurai & Matuo, 1957), although the fungus occurs on a wide variety of other hosts over a wide geographical range. Such differences could be due to environmental requirements not being met in certain regions, but the situation in *F. solani* f. sp. *cucurbitae* provides another explanation. Here some self-fertile forms occur, but perithecia of self-sterile forms

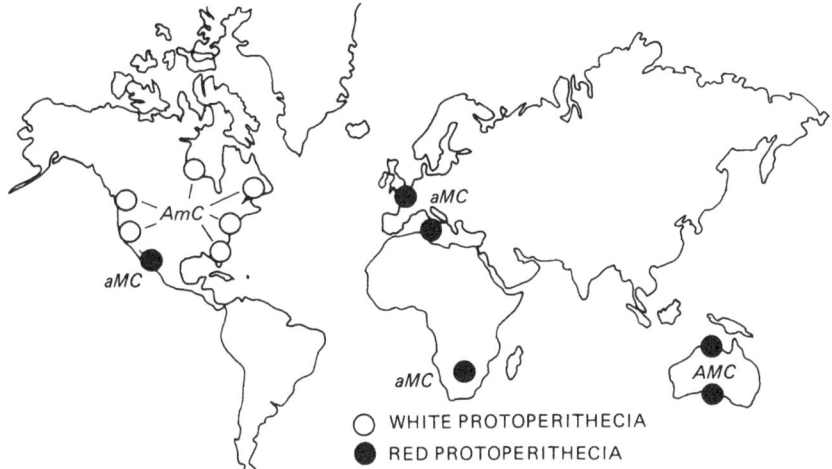

Fig. 3.1 World distribution of mating types (*A/a*), sexuality (*M/m* and *C/c*) and perithecial colour – red or white – in isolates of *F. solani* f. sp. *cucurbitae*, Race 2 received at Berkeley, California between 1945 and 1975. (Based on Snyder *et al.*, 1975.)

are unknown in nature although readily produced in the laboratory. This is because, in nature, fully complementary strains are widely separated geographically (Fig. 3.1) and have apparently only been brought together by man (Snyder *et al.*, 1975).

According to Gordon (1954), a superficially similar situation occurs in *F. sulphureum* Schlecht. (= *Gibberella cyanogena* (Desm.) Sacc.) but here most of the isolates are of *a* mating-type only. How such conditions have arisen is not known; one possibility is that widespread dispersal of single clones occurs, possibly through the agency of man. The fact that, in the U.S.A. and Canada, all natural isolates save one are of the same mating type, are male-deficient and carry the mutant allele for white perithecia (a combination not so far found elsewhere) supports the view that this population represents a single introduction. It also suggests that sexual reproduction may not be of great selective advantage to *F. solani* f. sp. *cucurbitae* which is therefore, in effect, an asexual pathogen throughout its range.

The genetic basis of reproduction in self-sterile forms is now clearly understood. Normal strains carry one of a pair of allelomorphic mating-type factors *A/a*. Each develops protoperithecia with trichogynes, microconidia and sporodochia with macroconidia. Sexual reproduction is effected by the activation of a conidium by a diffusate from a compatible mycelium. The conidium is then capable of attracting a trichogyne. Fusion is followed by migration of a 'male' nucleus to the ascogonium in the protoperithecium, which then develops into a perithecium with asci (Bistis & Georgopoulos, 1979). Female-deficient mutants ($C \rightarrow c$), lacking protoperithecia and sporodochia, occur more commonly than male-deficient mutants ($M \rightarrow m$). These (Cm) sometimes produce fewer microconidia than normal, hermaphrodite (CM) or female-deficient (cM) strains but, in any event, they cannot function as spermatising agents (Hansen & Snyder, 1946; El-Ani, 1954, 1956; Baker, 1956). Their conidia can be activated, attract and fuse with trichogynes but, thereafter, reproduction is blocked intracellularly in an unknown manner (Bistis & Georgopoulos, 1979). The loci concerned are about 18 units apart, on opposite sides of the centromere of a different chromosome from that carrying the A/a mating-type locus (Georgopoulos, 1963). Hence, rare 'neutral', i.e. sterile, cross-over types (cm) can arise from crosses between the deficient mutants: $Cm \times cM \rightarrow CM$ and cm (neutral). Such sexuality deficient mutants are now known to occur in several formae speciales both of *F. solani* and *F. moniliforme* (Matuo & Snyder, 1973; Chang & Sun, 1975).

Finally, it should be said that when viable ascospores are formed they are unordered. Hence genetic analysis has to be carried out either by the use of unordered tetrads or by random ascospore analysis. Both methods have been employed (e.g. tetrads: El-Ani, 1954; Georgopoulos, 1963; Tegtmeier & Van Etten, 1982*a*; random ascospores: Dimock, 1973*a, b, c*; Snyder, 1940; Van Etten, 1978).

The main snags of such procedures are the necessity of washing adherent conidia from the perithecia and the relatively small size of the ascospores, which makes them difficult to distinguish from germinating microconidia. Controlled crosses in self-fertile *Nectria haematococca* can now be achieved using appropriate combinations of self-sterility mutants (Babai-Ahary, Daboussi-Bareyre & Parisot, 1982).

Heterokaryosis and parasexuality

It is not clear how far recombination, either at the nuclear level, i.e. heterokaryosis, or at the chromosomal level through mitotic recombination, occurs in self-sterile strains of sexually reproducing species or in strictly anamorphic species such as *F. oxysporum*. Such evidence as is available is only sufficient to suggest that heterokaryosis and parasexuality could occur but is insufficient to assess its significance in nature.

Heterokaryosis. There is only one report of the detection of stable heterokaryosis in nature, in *F. moniliforme*. Ming, Lin & Yu (1966) found that 19 isolates from various places in China were mostly homokaryotic but some were heterkaryons, being composed of two, or rarely three, prototrophic homokaryotic types. Those latter, when propagated from uninucleate microconidia, differed genotypically in pigmentation, being purple, red or white on Leonian's medium; they also differed in pathogenicity and in their gibberellin production as 500:250:5 respectively. The white form was virtually non-pathogenic, the purple most, and the red intermediate. One tri-heterokaryotic isolate from nature was studied in detail. It exhibited all the colours of its component homokaryotic types in different parts of the medium at different times, but its gibberellin production and pathogenicity were, on average, intermediate between those of the purple and red types and it was stable. It proved possible to re-synthesise a stable heterokaryon through anastomoses between its component homokaryons.

Buxton (1954) obtained similar results experimentally. He allowed mixtures of microconidia of *F. oxysporum* f. sp. *gladioli* each representing different, naturally occurring morphological and colour variants to germinate together and, through the anastomoses between germ tubes, to develop heterokaryons. These presented a mosaic growth of homo- and heterokaryotic regions from which different nuclear types segregated in the conidia. Since the majority of fusaria studied (Punithalingam, 1975) have a multi-nucleate apical cell, the hyphal cells being uni-, or rarely, binucleate and all conidial nuclei being derived from one parental nucleus in each phialide, such results are not expected. The persistence of such a heterokaryon must depend upon the maintenance of the heterokaryotic nature of the multinucleate apical cell, the frequency of anastomoses in the older parts and the prevalance in older regions of heterokaryotic multinucleate cells. Indeed, in *F. oxysporum* f. sp. *callistephi* heterokaryosis must be confined to the older regions since the apical cell is uninucleate (Hoffman, 1966*a*). Segregation occurs either at conidium formation, or by the chance appearance of a homokaryotic apical cell – uni- or multinucleate – which gives rise to a homokaryotic sector. In *F. moniliforme* in culture the apical cell is usually 2, 3 or 4-nucleate (the hyphal cells normally being uninucleate), although up to 5-nucleate cells were seen (Punithalingam, 1975). This makes the reported stability of its heterokaryons even more surprising when compared with *F. oxysporum* f. sp. *gladioli*, with around 7 nuclei per apical cell, or f. sp. *callistephi* with 1, or 1.6–1.8 nuclei per hyphal cell. Nevertheless, it should be noted that Hoffmann (1964) has provided evidence that in culture the nuclear number can be affected by temperature, the C/N ratio and pH value of the medium.

Experimental studies of heterokaryosis in culture have usually employed auxotrophic mutants but, if the heterokaryons grew at all on unsupplemented medium, the growth was usually feeble, e.g. *F. oxysporum* f. sp. *callistephi* (Hoffmann, 1966*b*, 1967), *F. oxysporum* f. sp. *lycopersici* (Sanchez, Leary & Endo, 1976), and between various f. sp. of *F. oxysporum* (Garber, Wyttenbach & Dhillon, 1961). A common feature in many of the experimental situations studied was the unequal nuclear ratios detected in putative heterokaryons and their differential stability with changes in the composition of the medium (Buxton, 1954; Sanchez, Leary & Endo, 1976). By contrast, Buxton (1956) claimed to have produced balanced heterokaryons between different auxotrophic mutant strains of *F. oxysporum* f. sp. *pisi* Race 1, although he gives no quantitative data save on the disease ratings. Heterokaryons showing growth can also be forced between auxotrophs of self-fertile *Nectria haematococca* (= *F. solani*). Despite the presence of multinucleate cells (range 1–8 nuclei per hyphal cell) in the mycelium, when this was macerated only 0.5–1.5% of the hyphal fragments proved to be heterokaryotic. This suggests that in this fungus the heterokaryon exists as a mosaic, as also seems to be the case in *F. moniliforme* and *F. oxysporum* f. sp. *gladioli*. Clearly further investigations of heterokaryosis are desirable in different species of *Fusarium*.

Parasexuality. Coupled with experimental investigations of heterokaryosis have been investigations of the parasexual process. Prototrophic uninucleate microconidia, i.e. putative diploids, have been isolated very infrequently, e.g. 3 in 10^8, or 5–40 in 10^7, in *F. oxysporum* f. sp. *pisi* (Buxton, 1956; Tuveson & Garber, 1961), and 20 in 3×10^7 in f. sp. *cubense* (Buxton, 1962). However, in addition, in both *F. oxysporum* f. sp. *callistephi* (Hoffmann, 1967) and f. sp. *vasinfectum* (Ahamed & Shanmugasundaram, 1972) recombinants, although not diploid ones, were detected. These data, taken with the preceding discussion on heterokaryosis, suggest that if transient heterokaryosis can occur in some fusaria it provides an opportunity for nuclear fusion. Recombinants have also been detected in some cases, e.g. *F. oxysporum* f. sp. *pisi* and *cubense* (Buxton, 1956, 1962), f. sp. *callistephi* (Hoffman, 1967, 1968) and f. sp. *vasinfectum* (Ahamed & Shanmugasundaram, 1972); however they have only been observed in any quantity in f. sp. *cubense*

(21/3036 microconidial isolates , i.e. 1 in 145). There exists, therefore, suggestive evidence for the parasexual cycle in *Fusarium*, but the whole subject needs careful re-examination. Nevertheless, mitotic recombination is evidently a potential tool for the genetic analysis of fusaria.

Pathogenicity

Pathogenicity is the outcome of a complex interaction in time between host and pathogen, each potentially variable and in a changing environment. Nevertheless, it is convenient to distinguish between the host specificity of the pathogen and the severity of disease which it provokes in a single host or in a number of similar ones. Some fusaria are wholly saprophytic. Others, in addition to their saprophytic potential, also range from being weakly to highly pathogenic; none, however, are obligate parasites. They may be general, wide-ranging pathogens, or confined to a single species – even a single cultivar; furthermore they may be pathogenic in one evironment and saprophytic in another.

The terms 'physiological races' and formae speciales are used to describe the degree of host specificity. Some progress has been made in the genetical analysis of the origins and status of members of these categories, which may differ within and between species of *Fusarium*.

Host specificity

Physiological races. A physiological race is most satisfactorily defined in terms of the resistance genes in the host which it matches and, on Flor's (1956) gene-for-gene theory, this implies a comparable number of avirulence/virulence genes in the fungus. For n resistance genes in a host there are 2^n physiological races. Thus there are two resistance genes in the tomato, *R1* and *R2*, which confer resistance to different isolates of *F. oxysporum* f. sp. *lycopersici*. Potentially, four kinds of isolate can be recognised: one incapable of matching either resistance gene (*Av1 Av2*), two capable of matching one (*av1 Av2*) or the other (*Av1 av2*), and a fourth able to match both (*av1 av2*). If a third resistance gene were discovered there would, therefore, be eight potential races, i.e. 2^3, and so on. The genetic designations used here imply that virulence (*av*) is recessive and avirulence (*Av*) dominant. This is based on the situation found in fungi such as rusts which have been studied extensively and where, in addition, such genes are usually unlinked. A new race can often be equated to a mutation from avirulence to virulence.

Little is known about the spontaneous origin of new races of *Fusarium* in nature save that they arise irregularly, unpredictably and appear to be discrete events. They can also be induced by mutagens. Bouhot (1970, 1973, 1981, Bouhot & Louvet, 1971) has studied *F. oxysporum* f. sp. *melonis* which is confined to melon (*Cucumis melo*) as host. Three differential hosts carrying either no resistance genes (universal suscept: *fom 0, fom 1, fom 2*), *Fom 1*, or *Fom 2* (Risser & Mas, 1965; Risser, 1973) were employed and four of the potential pathogenic races, R_0, R_1, R_2 and $R_{1.2}$, are known from nature in Europe (Table 3.4).

Microconidia from each race were treated separately with nitrosoguanidine and a selection of treated isolates was tested on the three differential hosts. It can be seen (Table 3.4) that new races were produced as the result of one, or possibly two, independent mutations from avirulence to virulence, or *vice versa*. Some mutants represented races not identified from nature, the commonest being those which had lost their virulence entirely. Such isolates, unless morphologically distinct, would be indistinguishable from the many saprophytic isolates of *F. oxysporum* which occur in the soil. The mutants were maintained in sterile soil, in some cases for 7 years, and tested annually for stability. Some were quite stable but others either reverted to the parental type or exhibited a fluctuating phenotype. Thus in successive years a mutant from Race 1 (*av0 av1 Av2*) varied as follows:

Year 1:   *av0   av1   Av2*

Year 2:   ***Av0   Av1***   *Av2*

Years 3–5:   ***av0   av1   av2***

and this was stable for 2 more years. Similarly a mutant from Race 0 (*av0 Av1 Av2*) fluctuated as follows:

Year 1:   *av0   **av1   av2***

Year 2:   *av0   av1   **Av2***

Year 3:   *av0   av1   Av2*

Year 4:   *av0   **Av1**   Av2*

Year 5:   *av0   **av1**   Av2*

(Changed loci are shown in bold italic.)

Table 3.4. *Progeny of three different physiological races of* Fusarium oxysporum *f. sp.* melonis *after treatment with nitrosoguanidine*

	Progeny from treated Race				
	R_0	R_1	$R_{1.2}$		
1. Races known from France				*Totals*	
R_0 *av0 Av1 Av2*	(8)*	3	2	Parental recovered	29
R_1 *av0 av1 Av2*	1	(8)	0	Pathogenic races	24
R_2 *av0 Av1 av2*	0	2	1	Avirulent races	27
$R_{1.2}$ *av0 av1 av2*	2	1	(13)		
2. Races not known in France				*Total changes at loci in progeny*	
Av0 Av1 Av2	5+	8	14	Unchanged	87
Av0 av1 Av2	1	3	2	Virulent to avirulent	85
Av0 Av1 av2	2	1	0	Avirulent to virulent	24
Av0 av1 av2	0	0	3		

Based on Bouhot, 1981.
* Parental types in parentheses.
+ Completely avirulent types in italic.

The reason for this instability is not known. All that can be suggested is that the mutagen used is believed to promote base pair transitions:

$$
\begin{array}{cc}
A & G \\
|| \rightleftharpoons ||| \\
T & C
\end{array}
$$

and such substitutions are characteristically spontaneously reversible at low frequency (Fincham, Day & Radford, 1979).

Earlier, support for mutation as a basis of changed race specificity had been provided by Buxton (1956, 1962) for *F. oxysporum* f. sp. *pisi* and f. sp. *cubense*. After UV irradiation the virulence of Races 1 and 2 of f. sp. *pisi* to three differential pea cultivars was changed and a heterokaryon between these mutants exhibited the phenotype which would be expected if avirulence were dominant to virulence.

Unambiguous demonstrations of mutational change from virulence to avirulence and *vice versa* are uncommon in fungi (Day, 1974) but such data as are available are characterised by several extraordinarily high mutation rates as in Bouhot's and Buxton's data. It is, therefore, surprising that there is little evidence for physiological races in species other than *F. oxysporum*, and even here they are only known in 18 out of 74 formae speciales. Whether their absence reflects the inability of investigators to discover and test suitable differential hosts, or whether it is indeed the case that races have not developed in some formae or species, is not known. Data from *F. solani* f. sp. *pisi* suggest that even in a forma with a wide host range the latter is the case (Van Etten, 1978). If this represents the true picture, it will be of importance to discover the reasons for it. They could, of course, lie in the genetical architecture which determines virulence of the pathogen – a matter of some importance for resistance breeding.

Despite the ease with which Bouhot and Buxton achieved mutations to changed race specificity and the records of apparently similar spontaneous changes, many races are remarkably stable. Indeed, even when experimental attempts have been made to select rare spontaneous mutants by constantly exposing a resistant cultivar to massive, avirulent microconidial inocula for several generations, a change in virulence has not been detected, e.g. *F. oxysporum* f. sp. *pisi* Race 2. Such treatment can, however, alter the virulence of the isolate to suceptible cultivars quite rapidly (Asher & Burnett, unpublished). Such observations cast great doubt on Buxton's (1958) claim that exposure of a strain of Race 1 derived from a microconidial isolate could, if incubated for 14 days in the root exudate of the pea cultivar Alaska, acquire the ability to infect it, i.e. acquire a Race 2 characteristic. It seems even less likely, since he claimed that this change was cytoplasmic. If such changes occurred in the rhizosphere, then the development of new races in nature would surely be more frequent!

Formae speciales. These are morphologically similar strains characterised by their adaptation to different hosts and, preferably, named after the host species. However, the degree of restriction to a single host varies. For example in *F. oxysporum* the range extends from a single species to a wide range of unrelated hosts, viz.:

f. sp. *lini* restricted to flax (*Linum usitatissimum*)

f. sp. *lycopersici* restricted to the genus *Lycopersicum*

f. sp. *conglutinans* restricted to Cruciferae

f. sp. *vasinfectum* found on *Cajanus*, *Coffea*, *Glycine*, *Gossypium*, *Hevea*, *Hibiscus*, *Medicago*, *Ricinus*, *Solanum*, and *Vigna*.

Armstrong & Armstrong (1966, 1975) have developed a concept of primary and secondary hosts for forms of *F. oxysporum* which are not specific in pathogenicity to a single host. They suggest that common genes for pathogenicity must exist in those formae speciales which can exist on a common host. Although this may be a plausible speculation, very little is yet known about the genetics of such adaptations. They have also demonstrated that fusaria can infect plants without causing external symptoms of disease. These, therefore, act as symptomless carriers – e.g. *F. oxysporum* f. sp. *vasinfectum* derived from cotton is symptomless in the Mexican Clover (*Richardia scabra*) (Armstrong & Armstrong, 1948). If this phenomenon is widespread, fusaria may have far wider host ranges than detected at present.

At one extreme the basis of formae speciales is not dissimilar to that described for physiological races; at the other extreme it involves populations exhibiting total genetic isolation. Bouhot (1981) investigated five isolates of *F. oxysporum*, a saprophytic isolate, an isolate of f. sp. *gladioli*, and three isolates confined to different members of the Cucurbitaceae, namely, f. sp. *niveum*, confined to watermelon (*Citrullus lanatus*), f. sp. *melonis*, confined to melons (*Cucumis melo*) and f. sp. *cucumerinum* confined to cucumber (*Cucumis sativus*). Incidentally, it appears that *F. oxysporum* cannot apparently infect marrow (*Cucurbita pepo*). When natural isolates are used there is some slight cross-infection by the formae:

   (1) f. sp. *niveum* will attack seedlings of *Cucumis* but not the adult
   plants

(2) f. sp. *melonis* will attack seedlings of *Citrullus* but not the adult plants

(3) f. sp. *cucumerinum* will neither attack *Cucumis* nor *Citrullus*. Bouhot employed the same mutagen in the same manner as was described earlier (p. 52) to the saprophytic form, to f. sp. *gladioli* and to f. sp. *niveum* and then inoculated the three cucurbitaceous hosts with treated progeny. With the first two, no successful inoculations were detected but with f. sp. *niveum* one isolate lost the ability to infect *Citrullus* and acquired the ability to attack adult *Cucumis*. This new behaviour was retained throughout a two-year period and, by testing it against three differential melon cultivars employed to classify f. sp. *melonis*, it was demonstrated that the mutant had acquired the pathogenic attributes of Race 0 of this forma specialis. Clearly the differences between f. spp. *niveum* and *melonis* are not great and it seems possible that the number of genetic differences between them are sufficiently small – in respect of pathogenic behaviour – to be overcome by a combination of a few mutational events in a highly selective situation. Bouhot's further claim that from isolates of f. sp. *cucumerinum* inoculated on *Cucumis* and *Citrullus* all three formae speciales and intermediates between them could be isolated seems improbable. Clearly the work should be repeated, preferably with more genetic markers in the strains employed.

It has also been claimed that heterokaryons can be forced between auxotrophic mutants of different formae speciales (Dhillon, Garber & Wyttenbach, 1961; Garber, Wyttenbach & Dhillon, 1961). Their growth, however, was slight and the evidence largely circumstantial. Moreover, some of these cultures were derived from a remarkable, doubly-auxotrophic morphological mutant detected after UV-irradiation of *F. oxysporum* f. sp. *pisi* by Buxton (1959). This was said to produce not only a presumably self-fertile, teleomorphic *Nectria haematococca* stage in culture but also formed heterokaryons with various formae speciales both of *F. oxysporum* and *F. solani* (Buxton, 1959; Buxton & Ward, 1962). It is true that some workers have contended that these species overlap and cannot be separated (e.g. Hildreth, 1958; Madoshing, 1964; Bolton & Donaldson, 1972) but experienced workers claim this not to be so (e.g. Kraft, Burke & Haglund, 1981). Buxton's claims have been strongly disputed, albeit somewhat uncritically, by several very experienced workers (Gordon, 1960; Snyder & Alexander, 1961; Reichle, Snyder & Matuo, 1964). However, the most telling objection to Buxton's results, which cannot now be tested directly since

Table 3.5. *Mating populations (M.P.), macroconidial types and former designation of populations of* Fusarium solani *from the U.S.A. and Japan*

Mating population	Macroconidial type	Original designation (f. sp.)
I	A	*cucurbitae* Race 1
II	A	*batatas*
III	A	*mori*
IV	A	*xanthophylli*
VII	A	*robiniae*
?*	A	*eumartii* (in part)
?	B	*eumartii* (in part)
V	B	*cucurbitae* Race 2
VI	B	*pisi*
?	C	*radicicola* (formerly Race 1)
?	D	*phaseoli*

Based on Matuo & Snyder, 1973.
* Indicates perithecial stage not obtained in these matings.

his cultures are no longer available, comes from the situation in *F. solani*. Matuo and co-workers have shown that many of the formae speciales of *F. solani* are incapable of mating with each other and, incidentally, also exhibit constant differences in their macroconidial morphology. It seems unlikely that heterokaryons could be forced between isolates between which matings are not possible, although the basis of the intersterility is unknown and could be extracellular.

Seven distinct 'mating populations' (M.P.s), perhaps better described as cryptic sibling species, are now known for *F. solani* (Matuo & Snyder, 1973: Table 3.5, Fig. 3.2). Compatible matings within each M.P. result in the production of perithecia apparently indistinguishable from those first produced in compatible crosses within f. sp. *cucurbitae* and designated *Nectria haematococca*.

The seven M.P.s transcend the boundaries both of physiological races and formae as usually defined in this species and call into question the values of these designations both for pathology and for understanding the biology of fusaria. For example, Sakurai & Matuo (1960) showed that *F. solani* f. sp. *radicicola* Race 2, pathogenic both to potato and mulberry, was completely inter-fertile with f. sp. *pisi* which is now known to have a wide host range (Matuo & Snyder, 1972; Van Etten, 1978). But f. sp. *radicicola* Race 1, also pathogenic to potato (where it

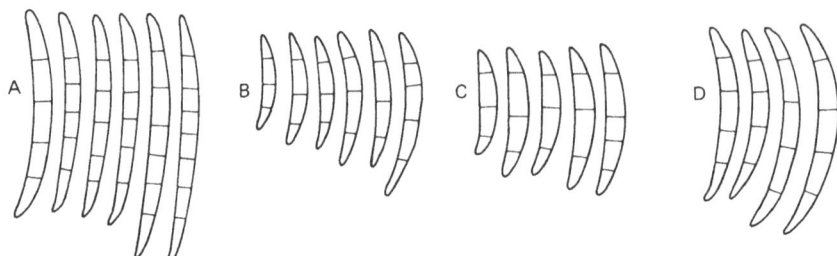

Fig. 3.2. Representative macroconidia of the four macroconidial classes found in mating populations of *Nectria haematococca* (=*Fusarium solani*). (After Matuo & Snyder, 1973.)

causes a tuber rot) but not to mulberry, will not cross with either of these formae, nor indeed with any other forms tested including f. sp. *eumartii*, the cause of basal rot in potato. This latter forma is itself separable into two groups on macroconidial characters comparable with those which distinguish other groups of M.P.s, e.g. f. sp. *pisi* from *radicicola* Race 1; however, it is not yet known whether these morphological groups can be crossed (Matuo & Snyder, 1973). By contrast, other formae which overlap and can be confused, e.g. f. sp. *pisi* and *phaseoli* (Yang & Hagedorn, 1965), are genetically isolated and morphologically separable. Even the two races of f. sp. *cucurbitae* turn out to be distinct and inter-sterile. Both infect cucurbits: Race 1 rots the cortex of the roots, stems and fruit, while Race 2 only causes a fruit rot. Lastly, it should be noted that the relationships of all these outbreeding M.P.s to the self-fertile strains, frequently saprophytic, or weakly parasitic and found in the tropics, Japan and the U.S.A., are unknown.

The *F. solani* pattern occurs in other sexually reproducing species. Several M.P.s are now known from *F. moniliforme**. One (**C**) is confined to rice, another (**B**) to sugarcane but a third (**A**) has been isolated from a pine, wheat, rye, and maize as well as from ants, aphids and ladybirds associated in the field with maize! Self-fertile strains are not known. Both self-fertile and outbreeding strains are known from *F. decemcellulare* (= *Calonectria rigidiuscula*). The former are saprophytic, the latter pathogenic to cocoa (*Theobroma*), but they can apparently be crossed although the viability of the progeny is unknown (Reichle & Snyder, 1964; Ford, Bourret & Snyder, 1967).

The basis of host specificity. There is a remarkable parallelism between

* See Hsieh *et al.*, 1977 and Kuhlman, 1982.

anamorphic species like *F. oxysporum* and sexually reproducing species like *F. solani*. The unconvincing evidence for heterokaryosis between formae in the former, and the existence of genetically-isolated mating populations in the latter, is coupled with a great morphological similarity between forms in each species. This suggests that such species of fusaria are in a state of dynamic speciation. Disruptive selection (Thoday, 1972), for which adaptation to different hosts or different parts of hosts provides ideal circumstances, appears to have led to different degrees of isolation in different species. As in so many other fungi (Burnett, 1975), cryptic sympatric speciation is common although ecological and geographical factors also play a role. Adaptation to a specific host is clearly not a sufficient cause for genetic isolation and the cases of *F. solani* f. sp. *pisi* (M.P. VI) or M.P. A of *F. moniliforme* also demonstrate that a wide host, or geographical, range can be maintained without apparent loss of inter-fertility.

These species provide opportunities for the genetical analysis of how populations react to competing selective pressures, either by the development of polymorphisms to match the heterogeneity of the environment, or by the subdivision of the initial population into isolated groups each adapted to a different niche within it. Here, there is a great opportunity for applying techniques such as somatic protoplast fusion (i.e. Kevei & Peberdy, 1977) to overcome barriers to heterokaryosis or breeding. The degree of genetic differentiation between populations is also amenable to measurement by gel-electrophoresis as has been done with other groups of closely related organisms (Ayala, 1975). It is clear that the pioneering studies of Reddy & Stahmann (1972) on the isoenzyme patterns of six formae of *F. oxysporum* or those of Royse & May (1982) with *Agaricus* could readily be developed and adapted to fusaria. Other techniques are also becoming available, e.g. the high-resolution crossed immuno-electrophoretic system used by Hornok (1980), to compare *Fusarium* spp., or the radio double-diffusion serological method of Iannelli *et al.* (1982) used with both formae and races of *F. oxysporum*. In these ways evolution in sexual and asexual forms could be compared.

Virulence

Different isolates may induce differing severities of disease on identical hosts: they differ in aggressiveness *sensu* Van der Plank (1968). The genetics of aggressiveness has received some attention and so has

Table 3.6. *Average disease rating of isolates of* Fusarium oxysporum *f. sp.* lycopersici *on cv. Bonny Best and Marglobe in relation to type of morphological variation*

Morphological type[+]	Disease Rating[*]	
	cv. Bonny Best	cv. Marglobe
Fluffy, aerial mycelium (*co*)	10.4	7.5
Raised sclerotial (*sp*)	8.7	6.3
Intermediate raised (*?ro*)	8.3	4.5
Intermedia appressed (*?slp*)	6.3	4.1
Slimy (*sh*)	4.7	3.0

Data of Wellmann & Blaisdell, 1940.
[+] The types are approximated to those listed in Table 3.1.
[*] Scale of disease rating 0–15, 15 being most severe.

the nature of the disease process. Those aspects of pathogenicity will now be considered.

Aggressiveness. Aggressiveness is commonly held to be correlated with morphology and with the origin of the strain under test.

The data of Table 3.6 illustrate a typical association between morphological variation of the type considered earlier (pp. 41–46) and aggressiveness. Many of the strains used by Wellman & Blaisdell (1941) had been in culture for long periods, several decades in some cases, and it is a common observation that prolonged culture is associated with both the development and replacement of one morphological variant by another and reduced pathogenicity (Miller, 1946). Nevertheless, Waite and Stover (1960) and later, Follin & Laville (1966) showed for *F. oxysporum* f. sp. *cubense* that virulence is not necessarily correlated with morphological form. Virulent strains differing in morphology can be isolated from the host and, if maintained under stable conditions in which senescence is selected against, each morphological type retains its original pathogenicity (Ming, Lin & Yu, 1966; Messiaen & Cassini, 1981). Using the same forma Buxton employed three strains of differing virulence isolated from cv. Gros Michel bananas, namely, B – very slightly virulent, yellowing type symptom; C – very highly virulent, yellowing type and D – moderately virulent, non-yellowing type. Page (1961) had already demonstrated that there were no stable cultural or physiological differences between these strains. Buxton employed auxotrophic and serological markers to identify each one and then prepared heterokaryons between them. Parasexual recombination took place and the recombinants showed two features. Firstly, the virulence-type was inherited intact and secondly, virulence-type recombined freely with the other markers. Taken together, the evidence suggests that morphological form (p. 42), virulence/avirulence (p. 51) and virulence-type all appear to be inherited independently as though determined by chromosomal genes. Thus the supposed correlations between aggressiveness and either morphological type, or the period maintained in culture, are likely to be fortuitous, arising instead from a combination of random mutation and the very different selective forces acting in culture as opposed to those acting in nature.

Evidence that aggressiveness is a property amenable to selection comes from observations on natural isolations. Matuo & Snyder (1972) and Van Etten (1978) have greatly extended the known host range of *F. solani* f. sp. *pisi* M.P. V1 by mating tests. Isolations by the former from pea, mulberry and ginseng (*Panax ginseng*) were tested for cross infectivity. Some 71% of the pea isolates showed the highest virulence rating when tested on pea cultivars compared with 45% of those from mulberry and none from ginseng. The weakly virulent isolates represented 24, 52 and 52% and the non-pathogenic, 5, 3 and 48% respectively from each host. The majority of the isolates most aggressive to pea were, therefore, derived from peas, with the least coming from ginseng. In contrast, when tested on mulberry, there were no significant differences in the mean virulence between isolates of different origin, most being of medium pathogenicity. On ginseng the results were similar to mulberry but here the isolates were only mildly pathogenic. Van Etten's data extended the range of host plants but he only tested comparative virulence on pea. Here too, isolates derived from peas were the most highly virulent, although some from unrelated hosts, e.g. potato, cottonwood (*Populus deltoides*), mulberry and sainfoin, were quite virulent (Table 3.7). Although a number of pea isolates when tested were weakly pathogenic or non-pathogenic to pea, the majority of isolates from other plants fell into these classes.

These data show clearly that there is a positive but incomplete correlation between the origin of an isolate and its aggressiveness. Similar, less well-quantified data exist for other formae with wide host ranges, e.g. *F. oxysporum* f. spp. *apii* and *vasinfectum* (Armstrong & Armstrong, 1967, 1975), and f. sp. *pisi* (Snyder, 1933).

How far selection for intra-racial virulence occurs in nature or can be achieved experimentally has not received much attention in fusaria. The

Table 3.7. *The average virulence of isolates of* F. solani *f. sp.* pisi *M.P.VI derived from different hosts to the same cultivars of pea*

Original host	Virulence rating*			
	11–8.1	8–5.1	5.0–1.6	1.5–0 (control)
Pisum sativum	5	6	3	7
Onobrychis viciifolia	—	1	1	—
Medicago sativa	—	—	—	3
Solanum tuberosum	—	1	1	—
Cicer arietinum	—	—	2	1
Morus nigra	—	1	3	1
Populus deltoides	—	1	—	2
Liriodendron tulipifera	—	—	2	1
Saprophyte	—	—	—	1

Based on Van Etten, 1978.
* Virulence rated on a scale 0–11, 11 being the most virulent.

necessary variability undoubtedly exists. For instance, Asher and Burnett (unpublished) demonstrated significant but not extensive variation in virulence amongst uninucleate microconidial isolates from a culture of *F. oxysporum* f. sp. *pisi* Race 2 after passage and re-isolation from a susceptible pea cv. Wisconsin Perfection. Moreover, when lines differing in virulence were separately maintained by single microconidial transfer through a range of differential hosts, significant variation continued to be exhibited at successive transfers by lines originally exhibiting a low level of virulence. The origin of this variation is not known. Reference has already been made to Buxton's claim of a cytoplasmic component of pathogenicity in *F. oxysporum*, and selection for aggressiveness in other fungi, e.g. *Phytophthora* (Caten, 1970) and *Rhynchosporium* (Habgood, 1973), has been attributed to extrachromosomal causes. But there seems no overriding reason to dismiss the combined results of gene mutation and selection. In Asher and Burnett's experiments less aggressive mutants would not be detected in any event and more aggressive mutant nuclei would be at a selective advantage in a pathogenic situation.

The situation would be well worth investigation in a species more readily amenable to genetic analysis such as *F. solani*.

Mechanisms of virulence and resistance. Several toxins (see Chapter 5) are produced by *Fusarium* spp. (Kern, 1972) and in many cases strains which differ in pathogenicity have been shown to differ in phytotoxin production *in vitro*. Marticin, for example, is produced *in vitro* by pathogenic strains of *F. solani* f. sp. *pisi* (150 mg/l after 4 days) whereas, under the same conditions, a weakly pathogenic strain produces none in the same time and very little indeed even after 15 days. Marticins have also been isolated from diseased plants at levels high enough to cause tissue damage. Similarly, pathogenic and non-pathogenic strains of *F. oxysporum* f. sp. *lycopersici* differ in their production of the non-specific toxins fusaric acid and lycomarasmin. On the other hand in none of these cases are the effects of the toxin unequivocally correlated with the disease symptoms found in nature and strains of f. sp. *lycopersici*, pathogenic in nature, have been found which are incapable of producing either fusaric acid or lycomarasmin. Clearly genetic analysis, comparable to that carried out in connexion with the host specific toxin victorin of *Helminthosporium victoriae* and the gene-for-gene relationship with resistance in the host, oats (*Avena*) (Scheffer & Yoder, 1972), is both possible and desirable but, to date, such studies have thrown little light on the nature of pathogenicity or resistance.

A more promising situation has been revealed by a combined physiological and genetical analysis of pathogenicity in *F. solani* f. sp. *pisi*. Highly pathogenic field isolates are found to be tolerant of the phytoalexin pisatin, produced by the pea, and capable of demethylating it; weakly pathogenic isolates lack this trait (Van Etten *et al.*, 1980; Tegtmeier & Van Etten, 1982*b*). Crosses between such isolates were either infertile, or produced progeny of one parental type only. This suggests some kind of lethal effect in the dikaryotic or diploid state. However, crosses between moderately pathogenic and weakly pathogenic isolates were fertile and tetrad analysis was possible; unfortunately the results were difficult to interpret because of the low percentage germination of the ascopores (about 50–80%). In most crosses the progeny segregated 1 : 1 pisatin tolerant : pisatin sensitive but some segregated 6 : 2, or 2 : 6, and one segregated 8 : 0. In almost every case pisatin tolerance was correlated with the ability to demethylate it and *vice versa*, and such progeny were always weakly- or non-pathogenic. It seems reasonable to conclude that at least two loci were segregating for demethylating ability. However, in one cross there was a range of tolerance and sensitivity to pisatin amongst the progeny and a few showing high sensitivity were also able to demethylate pisatin. Thus the situation may well be more complex but, particularly if ascospore germination can be improved, genetic analysis of pathogenicity in this pathogen appears possible. A parallel investigation of the relationship between phytoalexin production and resistance genes in the host is equally desirable using isolines (Smith & MacHardy, 1982).

Conclusions

It is clear that the fungi classified somewhat artificially on their teleomorphic state as *Fusarium* can provide material for a range of genetic investigations at the molecular, cellular and population levels. It will be noted that there is virtually no genetic data on metabolic pathways, hence the genetics of the remarkable range of secondary products which fusaria are known to produce is as yet unexplored. Attention has already been drawn to the possibilities for the genetic exploration of the mechanisms of reproduction, parasitism and virulence. Lastly, it will be apparent that the fusaria provide a wonderful range of material for the study of selection, population divergence and speciation, especially in its early stages.

References

Ahamed, N. M. M. & Shanmugasundaram, E. R. B. (1972). Biochemical and genetical studies in host–parasite relationships. Effect of heterokaryosis on the degree of virulence of nutritional mutants of *Fusarium vasinfectum*. *Phytopathologische Zeitschrift*, **75**, 349–59.

Armstrong G. M. & Armstrong, J K. (1948) Nonsusceptible hosts as carriers of wilt fusaria. *Phytopathology*, **38**, 808–26.

Armstrong, G. M. & Armstrong, J. K. (1966). *Fusarium oxysporum* f. *cassiae* form. nov., causal agent of wilt of *Cassia tora* and other plants. *Phytopathology*, **56**, 699–701.

Armstrong, G. M. & Armstrong, J. K. (1967). The celery-wilt *Fusarium* causes wilt of garden pea. *Plant Disease Reporter*, **51**, 888–92.

Armstrong, G. M. & Armstrong, J. K. (1975). Reflections on the wilt fusaria. *Annual Review of Phytopathology*, **13**, 95–103.

Armstrong, G. M. & Armstrong, J. K. (1980). Formae speciales and races of *Fusarium oxysporum* causing wilt diseases. In *Fusarium, Diseases, Biology, and Taxonomy*, ed. P. E. Nelson, T. A. Toussoun & R. J. Cook, pp. 391–9. University Park & London: The Pennsylvania State University Press.

Ayala, F. J. (1975). Genetic differentiation during the speciation process. *Evolutionary Biology*, **8**, 1–78.

Babai-Ahary, A., Daboussi-Bareyre, M J. & Parisot, D. (1982). Isolation and genetic analysis of self-sterility and perithecial pigment mutants in a homothallic isolate of *Nectria haematococca*. *Canadian Journal of Botany*, **60**, 79–84.

Baker, R. (1956). Fertilizing ability of males and hermaphrodites in *Hypomyces solani* f. *cucurbitae*. *Phytopathology*, **46**, 644–9.

Bareyre, M. J. & Laillier-Rousseau, D. (1972). Propriétés de nouveaux mutants affectant les variations morphologiques chez le *Nectria haematococca*. *Compte rendu hebdomadaire des séances de l'Académie des Sciences, Paris, Serie D*, **274**, 3614–15.

Bistis, G. N. & Georgopoulos, S. G. (1979). Some aspects of sexual reproduction in *Nectria haematococca* var. *cucurbitae*. *Mycologia*, **71**, 127–43.

Bolton, A. T. & Donaldson, A. G. (1972). Variability in *Fusarium solani* f. *pisi* and *F. oxysporum* f. *pisi*. *Canadian Journal of Plant Science*, **52**, 189–96.

Booth, C. (1971). *The Genus Fusarium*. Kew: Commonwealth Mycological Institute.

Booth, C. (1975). The present status of *Fusarium* taxonomy. *Annual Review of Phytopathology*, **13**, 83–93.

Bouhot, D. (1970). Variations induités du pouvoir pathogène chez *Fusarium oxysporum* f. sp. *melonis*. *Annali Academia Scientia Fennica, A, IV Biologica*, **168**, 25–7.

Bouhot, D. (1973). Some studies on the origin of races and formae speciales in *Fusarium oxysporum* by use of nitrosoguanidine mutants. *Abstract*, 2nd International Congress of Plant Pathology, Minneapolis, Minnesota.

Bouhot, D. (1981). Some aspects of the pathogenic potential in *formae speciales* and races of *Fusarium oxysporum* on Cucurbitaceae. In *Fusarium, Diseases, Biology, and Taxonomy*, ed. P. E. Nelson, T. A. Toussoun & R. J. Cook, pp. 318–26. University Park & London: The Pennsylvania State University Press.

Bouhot, D. & Louvet, J. (1971). Some observations and experiments on the origin of *Fusarium oxysporum* f. sp. *melonis* races in France. *International Symposium on Pathological Wilts of Plants*, Madras, 18–25.

Bouvier, J. & Laville, E. (1970). Origine et fonction des inducteurs d'états différenciés chez deux Ascomycètes. *Physiologie Végétale*, **8**, 361–74.

Brown, W. (1926). Studies in the genus *Fusarium IV*. On the occurrence of saltations. *Annals of Botany*, **40**, 223–43.

Brown, W. & Horne, A. S. (1926). Studies in the Genus *Fusarium III*. An analysis of factors which determine certain microscopic features of *Fusarium* strains. *Annals of Botany*, **40**, 203–21.

Burnett, J. H. (1975). *Mycogenetics*. London: John Wiley & Sons.

Buxton, E. W. (1954). Heterokaryosis and variability in *Fusarium oxysporum* f. *gladioli* (Snyder & Hansen). *Journal of General Microbiology*, **10**, 71–84.

Buxton, E. W. (1956). Heterokaryosis and parasexual recombination in pathogenic strains of *Fusarium oxysporum*. *Journal of General Microbiology*, **15**, 133–9.

Buxton, E. W. (1958). A change of pathogenic race in *Fusarium oxysporum* f. *pisi* induced by root exudate from a resistant host. *Nature*, **181**, 1222–4.

Buxton, E. W. (1959). Production of a perfect stage in a nutritionally deficient mutant of pathogenic *Fusarium oxysporum* after ultra-violet irradiation. *Nature*, **184**, 1258.

Buxton, E. W. (1962). Parasexual recombination in the Banana-wilt fungus. *Transactions of the British Mycological Society*, **45**, 274–9.

Buxton, E. W. & Ward, V. (1962). Genetic relationships between pathogenic strains of *Fusarium oxysporum*, *Fusarium solani* and an isolate of *Nectria haematococca*. *Transactions of the British Mycological Society*, **45**, 261–73.

Carlile, M. J. (1956). A study of factors influencing non-genetic variation in a strain of *Fusarium oxysporum*. *Journal of General Microbiology*, **14**, 643–54.

Caten, C. E. (1970). Spontaneous variation of single isolates of *Phytophthora infestans*. II. Pathogenic variation. *Canadian Journal of Botany*, **48**, 897–905.

Chang, I. C. & Sun, S. K. (1975). The perfect stage of *Fusarium moniliforme*. *Journal of Agricultural Research*, China, **24**, 11–20.

Daboussi-Bareyre, M. J. (1976). Synthèse et migration de l'information morphogénétique chez le *Nectria haematococca* (Berk. et Br.) Wr. I Etude chez un mutant conditionnel. *Physiologie Végétale*, **14**, 517–32.

Daboussi-Bareyre, M. J. (1977). Synthèse et migration de l'information morphogénétique chez le *Nectria haematococca* (Berk. et Br.) Wr. II. Etude des modalités de l'expression. *Physiologie Végétale*, **15**, 577–90.

Daboussi-Bareyre, M. J. (1980). Heterokaryosis in *Nectria haematococca. Journal of General Microbiology*, **116**, 425–33.

Daboussi-Bareyre, M. J., Laillier-Rousseau, D. & Parisot, D. (1979). Contrôle génétique de deux états différenciés de *Nectria haematococca. Canadian Journal of Botany*, **57**, 1161–73.

Day, P. R. (1974). *Genetics of Host Parasite Interaction*. San Francisco: W. H. Freeman & Co.

Dhillon, T. S., Garber, E. D. & Wyttenbach, E. G. (1961). Genetics of phytopathogenic fungi. VI. Heterocaryons involving *Gibberella fujikuroi* and formae of *Fusarium oxysporum. Canadian Journal of Botany*, **39**, 785–92.

Dimock, A. W. (1937*a*). Observations on sexual relations in *Hypomyces ipomoeae. Mycologia*, **29**, 116–27.

Dimock, A. W. (1937*b*). Hybridisation studies on a zinc-induced variant of *Hypomyces ipomoeae. Mycologia*, **29**, 273–85.

Dimock, A. W. (1937*c*). Hybridisation experiments with natural variants of *Hypomyces ipomoeae. Bulletin of the Torrey Botanical Club*, **64**, 499–507.

El-Ani, A. S. (1954). The genetics of sex in *Hypomyces solani* f. *cucurbitae. American Journal of Botany*, **41**, 110–13.

El-Ani, A. S. (1956). Ascus development and nuclear behaviour in *Hypomyces solani* f. *cucurbitae. American Journal of Botany*, **43**, 769–78.

Fincham, J. R. S., Day, P. R. & Radford, A. (1979). *Fungal genetics*, 4th edn. Oxford: Blackwell Scientific Publications.

Flor, H. H. (1956). The complementary genic systems in flax and flax rust. *Advances in Genetics*, **8**, 29–54.

Follin, J. C. & Laville, E. (1966). Variations chez le *Fusarium oxysporum* f. *cubense* (Agent causal de la maladie de Panama du bananier). *Fruits*, **21**, 261–8.

Ford, E. J., Bourret, J. A. & Snyder, W. C. (1967). Biological specialization in *Calonectria (Fusarium) rigidiuscula* in relation to green-point gall of cocoa. *Phytopathology*, **57**, 710–12.

Francis, R. G. & Burgess, L. W. (1977). Characteristics of two populations of *Fusarium roseum* 'Graminearum' in Eastern Australia. *Transactions of the British Mycological Society*, **68**, 421–7.

Garber, E. C., Wyttenbach, E. G. & Dhillon, T. S. (1961). Genetics of phytopathogenic fungi. V Heterocaryons involving formae of *Fusarium oxysporum. American Journal of Botany*, **48**, 325–9.

Georgopoulos, S. G. (1963). Genetic markers and linkage relationships in *Hypomyces solani* f. *cucurbitae. Canadian Journal of Botany*, **41**, 649–59.

Gordon, W. L. (1954). Geographical distribution of mating types in *Gibberella cyanogena* (Desm.) Sacc. *Nature*, **173**, 505.

Gordon, W. L. (1960). Is *Nectria haematococca* the perfect stage of *Fusarium oxysporum*? *Nature*, **183**, 903.

Habgood, R. M. (1973). Variation in *Rhynchosporium secalis. Transactions of the British Mycological Society*, **61**, 41–7.

Hansen, H. N. & Snyder, W. C. (1946). Inheritance of sex in fungi. *Proceedings of the National Academy of Sciences*, Washington, **32**, 272–3.

Harter, L. L. (1939). Influences of light on the length of the conidia in certain species of *Fusarium. American Journal of Botany*, **26**, 234–43.

Hildreth, R. C. (1958). Genetic variation and variability of *F. solani* f. *pisi* and *F. oxysporum* f. *pisi* Race 2. *Dissertation Abstracts*, **18**, 1196.

Hoffman, G. M. (1964). Untersuchungen über die Kernverhältmisse bei *Fusarium oxysporum* f. *callistephi*. *Archiv für Mikrobiologie*, **49**, 51–63.

Hoffman, G. M. (1966a). Untersuchungen über die Heterokaryosebildung und den Parasexualcyclus bei *Fusarium oxysporum*. I. Anastomosenbildung im Mycel und Kernverhältnisse bei der Conidienentwicklung. *Archiv für Mikrobiologie*, **53**, 336–47.

Hoffmann, G. M (1966b). Untersuchungen über die Heterokaryosebildung und den Parasexualcyclus bei *Fusarium oxysporum*. II. Gewinnung und Identifizierung auxotrophen Mutanten. *Archiv für Mikrobiologie*, **53**, 348–57.

Hoffman, G. M. (1967). Untersuchungen über die Heterokaryosebildung und den Parasexualcyclus bei *Fusarium oxysporum*. III. Paarungsversuche mit auxotrophen Mutanten von *Fusarium oxysporum* f. *callistephi*. *Archiv für Mikrobiologie*, **56**, 40–59.

Hoffmann, G. M. (1968). Kernverhältnisse bei pflanzenpathogenen imperfekten Pilzen, inbesondere Arten der Gattung *Fusarium*. *Zentralblatt für Bakteriologie, Parasitekunde und Infektionskrankheiten, Abteilung 2*, **122**, 405–519.

Horne, A. S. & Mitter, J. H. (1927). Studies in the genus *Fusarium*. V factors determining septation and other features in the section Discolor. *Annals of Botany*, **41**, 519–47.

Hornok, L. (1980). Serotaxonomy of *Fusarium* species of the sections Gibbosum and Discolor. *Transactions of the British Mycological Society*, **74**, 73–8.

Hsieh, W. H., Smith, S. N. & Snyder, W. C. (1977). Mating groups in *Fusarium moniliforme*. *Phytopathology*, **67**, 1041–3.

Hwang, Shuh-wei. (1948). Variability and perithecium production in a homothallic form of the fungus *Hypomyces solani*. *Farlowia*, **3**, 315–26.

Iannelli, D., Capparelli, R, R., Cristinzio, G., Marziano, F., Scala, F. & Noviello, C. (1982). Serological differentiation among *formae speciales* and physiological races of *Fusarium oxysporum*. *Mycologia*, **74**, 313–19.

Kern, H. (1972). Phytotoxins produced by Fusaria. In *Phytotoxins in plant disease*, ed. R. K. S. Wood, A. Ballis, & A. Graniti, pp. 35–48. London: Academic Press.

Kevei, F. & Peberdy, J. F. (1977). Interspecific hybridisation between *Aspergillus nidulans* and *Aspergillus rugulosus* by fusion of somatic protoplasts. *Journal of General Microbiology*, **102**, 255–62.

Kraft, J. M., Burke, D. W. & Haglund, W. A. (1981). *Fusarium* diseases of beans, peas and lentils. In *Fusarium, Diseases, Biology, and Taxonomy*, ed. P. E. Nelson, T. A. Toussoun & R. J. Cook, pp. 142–56. University Park & London: The Pennsylvania State University Press.

Kuhlman, E. G. (1982). Varieties of *Gibberella fujikuroi* with anamorphs in *Fusarium* section Liseola. *Mycologia*, **74**, 759–68.

Madoshing, C. (1964). A serological comparison of three *Fusarium* species. *Canadian Journal of Botany*, **42**, 1143–6.

Matuo, T. & Snyder, W. C. (1972). Host virulence and the *Hypomyces* stage of *Fusarium solani* f. sp. *pisi*. *Phytopathology*, **62**, 731–5.

Matuo, T. & Snyder, W. C. (1973). Use of morphology and mating populations in the identification of *formae speciales* in *Fusarium solani*. *Phytopathology*, **63**, 562–5.

Messiaen, C. M. & Cassini, R. (1981). Taxonomy of *Fusarium*. In *Fusarium, Diseases, Biology, and Taxonomy*, ed. P. E. Nelson, T. A. Toussoun & R. J. Cook, pp. 426–55. University Park & London: The Pennsylvania State University Press.

Miller, J. J (1946). Cultural and taxonomic studies on certain Fusaria. I Mutation in culture. *Canadian Journal of Research, C*, **24**, 188–212.

Ming, Y. N., Lin, P. C. & Yu, T. F. (1966). Heterokaryosis in *Fusarium fujikuroi* (Saw.) Wr. *Scientia Sinica*, **15**, 371–8.

Page, O. T. (1961). Variation in the banana-wilt pathogen *Fusarium oxysporum* f. *cubense*. *Canadian Journal of Botany*, **39**, 545–57.

Puhalla, J. E. (1981). Genetic considerations of the Genus *Fusarium*. In *Fusarium, Diseases, Biology, and Taxonomy*, ed. P. E. Nelson, T. A. Tousson & R. J. Cook, pp. 291–305. University Park & London: The Pennsylvania State University Press.

Punithalingam, E. (1975). Cytology of some *Fusarium* species. *Nova Hedwigia*, **26**, 275–304.

Reddy, M. N. & Stahmann, M. A. (1972). Isozyme patterns of *Fusarium* species and their significance in taxonomy. *Phytopathologische Zeitschrift*, **74**, 115–24.

Reichle, R. E. & Snyder, W. C. (1964). Heterothallism and ascospore numbers in *Calonectria rigidiuscula*. *Phytopathology*, **54**, 1297–9.

Reichle, R. E., Snyder, W. C., & Matuo, T. (1964). *Hypomyces* stage of *Fusarium solani* f. *pisi*. *Nature*, **203**, 664–5.

Risser, G. (1973). Etude de l'hérédite de la résistance du melon (*Cucumis melo*) aux races 1 et 2 de *Fusarium oxysporum* f. *melonis*. *Annales de l'Amelioration des Plantes*, **23**, 259–63.

Risser, G. & Mas, P. (1965). Mise en évidence de plusieurs races de *Fusarium oxysporum* f. *melonis*. *Annales de l'Amelioration des Plantes*, **15**, 405–8.

Royse, D. J. & May, B. (1982). Genetic relatedness and its application in the selective breeding of *Agaricus brunnescens*. *Mycologia*, **74**, 569–72.

Sakurai, Y. & Matuo, T. (1957). On a *Fusarium* disease of mulberry twigs caused by *Hypomyces solani* (Rke. et Berth.) Snyd et Hans. *Research Report of the Faculty of Textile and Sericulture, Shinshu University*, **7**, 18–24.

Sakurai, Y. & Matuo, T. (1960). Studies on the intraspecific groups in *Fusarium solani*. I On mating populations and morphologic groups in the species. *Research Report of the Faculty of Textile and Sericulture, Shinshu University*, **10**, 21–32.

Sanchez, L. E., Leary, J. V. & Endo, R. M. (1975). Chemical mutagenesis of *Fusarium oxysporum* f. sp. *lycopersici*: non-selected changes in pathogenicity of auxotrophic mutants. *Journal of General Microbiology*, **87**, 326–32.

Sanchez, L. E., Leary, J. V. & Endo, R. M. (1976). Heterokaryosis in *Fusarium oxysporum* f. sp. *lycopersici*. *Journal of General Microbiology*, **93**, 219–26.

Scheffer, R. P. & Yoder, O. C. (1972). Host-specific toxins and selective toxicity. In *Phytotoxins in Plant Diseases*, ed. R. K. S. Wood, A. Ballis & A. Graniti, pp. 251–72. London: Academic Press.

Sideris, C. P. (1925). The role of hydrogen-ion concentration on the development of pigment in fusaria. *Journal of Agricultural Research*, **30**, 1011–19.

Singh, U. P. (1973). Effect of acriflavine on UV-induced mutants of *Fusarium* species. *Mycopathologia et Mycologia applicata*, **50**, 183–93.

Smith, C. A. & MacHardy, W. E. (1982). The significance of tomatine in the host response of susceptible and resistant tomato isolines infected with two races of *Fusarium oxysporum* f. sp. *lycopersici*. *Phytopathology*, **72**, 415–19.

Smith, E. F. & Swingle, D. B. (1904). The dry rot of potatoes due to *Fusarium oxysporum*. *U.S. Department of Agriculture, Bureau of Plant Industry*, Bulletin **55**.

Snyder, W. C. (1933). A new vascular *Fusarium* disease of peas. *Science*, **77**, 327.

Snyder, W. C. (1940). White perithecia and the taxonomy of *Hypomyces ipomoeae*. *Mycologia*, **32**, 646–8.

Snyder, W. C. & Alexander, J. V. (1961). Perfect stages of *Fusarium oxysporum* and

Fusarium solani f. *pisi* still unknown. *Nature*, **189**, 596.

Snyder, W. C., Georgopoulos, S. G., Webster, R. K. & Smith, S. N. (1975). Sexuality and genetic behaviour in the fungus. *Hypomyces (Fusarium) solani* f. sp. *cucurbitae*. *Hilgardia*, **43**, 161–85.

Snyder, W. C. & Hansen, H. N. (1940). The species concept in *Fusarium*. *American Journal of Botany*, **27**, 64–7.

Snyder, W. C. & Hansen, H. N. (1941a). The species concept in *Fusarium* with reference to section Martiella. *American Journal of Botany*, **28**, 738–42.

Snyder, W. C. & Hansen, H. N. (1941b). The effect of light on taxonomic characters in *Fusarium*. *Mycologia*, **33**, 580–91.

Tegtmeier, K. J. & Van Etten, H. D. (1982a). Genetic studies on selected traits of *Nectria haematococca*. *Phytopathology*, **72**, 604–7.

Tegtmeier, K. J. & Van Etten, H. D. (1982b). The role of pisatin tolerance and degradation in the virulence of *Nectria haematococca* on peas: a genetic analysis. *Phytopathology*, **72**, 608–12.

Thoday, J. M. (1972). Disruptive selection. *Proceedings of the Royal Society of London, Series B*, **182**, 109–43.

Tuveson, R. W. & Garber, E. D. (1961). Genetics of phytopathogenic fungi. IV. Experimentally induced alterations in nuclear ratios of heterocaryons of *Fusarium oxysporum* f. *pisi*. *Genetics*, **46**, 485–92.

Van der Plank, J. E. (1968). *Disease Resistance in Plants*. New York: Academic Press.

Van Etten, H. D. (1978). Identification of additional habitats of *Nectria haematococca* mating population VI. *Phytopathology*, **68**, 1552–6.

Van Etten, H. D., Mathews, P. S., Tegtmeier, K. J. & Dietert, M. F. (1980). The association of pisatin tolerance and demethylation with virulence on Pea in *Nectria haematococca*. *Physiological Plant Pathology*, **16**, 257–68.

Waite, B. H. & Stover, R. H. (1960). Studies on *Fusarium* wilt of bananas. VI. Variability and the cultivar concept in *Fusarium oxysporum* f. *cubense*. *Canadian Journal of Botany*, **38**, 985–94.

Wellman, F. L. & Blaisdell, D. J. (1941). Pathogenic and cultural variation among single spore isolates from strains of the tomato-wilt *Fusarium*. *Phytopathology*, **31**, 103–20.

Yang, S. M. & Hagedorn, D. J. (1965). Pathogenicity studies with *Fusarium solani* from beans and peas. *Phytopathology*, **55**, 1085.

4

Fusarium and plant pathology: the reservoir of infection

D.PRICE

Glasshouse Crops Research Institute, Worthing Road, Rustington, Littlehampton, West Sussex BN16 3PU, UK

Introduction

Until about 30 years ago plant pathologists tended to look at diseases in isolation. The cause of a particular disease would be established and an appropriate control measure recommended, be it a fungicidal seed dressing or a spray. Occasionally, with intractable problems such as Panama disease, caused by *Fusarium oxysporum* f. sp. *cubense*, long-term breeding programmes for resistance were commenced. However, during the last three decades society has become more aware of its environment. Social consciences have developed and the indiscriminate use of land, and of agricultural chemicals, is increasingly questioned. Plant pathologists have also moved from a narrow view of disease to a recognition that the niche occupied by a pathogen in its environment, in the widest sense, is as important as the disease itself.

A good example of this is the changed appreciation of the significance of chlamydospores. A few decades ago the presence or absence of chlamydospores in culture was of importance only to taxonomists. Their existence in soil was suggested but not proven. Yet today a large number of pathologists see chlamydospores as the individual members of an infection reservoir; a reservoir of variable size. To understand the causes of variation, it is agreed, is to understand (and therefore control?) the disease.

Fusarium is an ubiquitous genus, of about 40 species of which more than half are parasitic upon green plants. Much of its success as a parasite can be attributed to its spread from infection foci. Yet it would be an unsuccessful pathogen if it were not able to exist actively away from its hosts. This point is important for two reasons. First, as chlamydospores have a limited life, crop rotations of suitable lengths would, in themselves, give good or even total control of a disease. Secondly, formae speciales have been found in circumstances that preclude the presence of the recognised host (Katan, 1971; Price, 1975a; Price & Linfield, unpublished).

Thus there is still much to be learned about *Fusarium* and its activities away from its recognised hosts. Yet even before experimental work begins the problem facing a pathologist is not the applied mycology but *Fusarium* taxonomy. The array of international taxonomic systems is impressive and although many are related there are pitfalls for the unwary unless one has a degree of familiarity with the major ones.

Booth (1971) lists 43 *Fusarium* species which may be divided into four main groups: plant pathogens (including mycoparasites), insect pathogens, saprophytes and soil inhabitants. A few species bridge the gap between groups, attacking both plants and insects or being able to live actively away from their host. Of the 43, 27 are pathogenic to green plants and amongst these are some of the most serious pathogens in world agriculture. The influence of these on crop yields or potential yields is enormous.

These pathogens may be put into one of three major groups: the wilts, caused by *F. oxysporum*; those attacking graminaceous plants, primarily *F. moniliforme, F. graminearum, F. avenaceum* and *F. culmorum*; and the root rots caused principally by *F. solani*. Each of these groups has factors in common with the others, e.g. method of survival in the absence of the host and possible genetic changes by sexual, parasexual or heterokaryotic means.

Little or no progress can be made in studying pathogenic species of *Fusarium* in the presence or absence of hosts unless the experimental techniques are good. Although useful work may start in the laboratory with *in vitro* tests, ultimately it is when experimenting in the usual environment of the pathogen that many investigations fail or give equivocal results. Immense problems have to be overcome if there is a need to differentiate between races and strains of the same species, especially when the difference is not an easily discernible taxonomic character. Thus, in random checks on isolates of *F. oxysporum* f. sp. *tulipae* some produced ethylene while others did not (Price, 1975b). Were these two strains of the same forma specialis? The difference may seem little to a taxonomist but is of great importance to the scientific adviser or grower since the effects of ethylene in bulb stores may be devastating, as low concentrations can lead to flower abnormalities (De Munk, 1971).

Equally important are pathogenicity tests where the results from the use of wound inoculations must be examined continuously because of seasonal physiological changes in host plant tissues. How valid are the results of inoculations using conidia or mycelial fragments when, in the natural habitat, chlamydospores are the pathogenic agent?

The sequence for studying *Fusarium* depends upon sampling systems and selective media. Although the use of rose bengal in a nutritious agar was very useful in its time the selective medium of Nash & Snyder (1962) must rank as one of the major advances in *Fusarium* pathology. Nash medium or a development of it is widely used and enables reasonably accurate comparisons of *Fusarium* soil flora in different parts of the world.

It is significant that of the 27 plant pathogenic species 18 form chlamydospores. Of the remainder eight can reproduce sexually, whilst with *F. arthrosporioides* and *F. poae* neither a perfect stage nor chlamydospores have yet been found. *F. sambucinum* has both chlamydospores and a perfect stage.

Although there are several effective fungicides used on a world-wide basis *Fusarium* diseases are as serious as ever and it seems that the primary source, the chlamydospore, has been relatively unaffected. This review, therefore, focuses on the chlamydospore. The literature on *Fusarium* pathology is vast and many hundreds of references and papers have been examined, though comparatively few of them are quoted here.

Ecological niches

In unamended soil, populations of *Fusarium* will exist as chlamydospores, and only if there are crop residues or plant roots present are there likely to be conidia or mycelial fragments (Snyder, 1969). It is ubiquitous as a pathogen because it has been recorded as such but there are many examples where it must have existed before its presence was displayed in this way. Colonised plant residues are often the remains of a parasitised plant: thus one speaks of soil plus residues. But what is the role of a forma specialis in soil before it had ever actively parasitised its recognised host? In the early 1950s, in the then Southern Cameroons, the Cameroons Development Corporation began breeding 'deli × dura' hybrids of the oil palm *Elaeis guinensis* using, as parents, palms that had been planted some 20 to 30 years earlier. These parents were sound, apparently healthy trees in blocks of apparently healthy palms. Many thousands of hybrid seedlings had been planted by 1959 in different parts of the Victoria Division. In 1959, in the space of a few weeks, some hybrid palms died in an area which had not been known to have previously supported any oil palms. They died from vascular wilt disease caused by *F. oxysporum* f. sp. *elaeidis*. Previously unrecorded in the Cameroons, the disease spread rapidly and the source of the original inoculum has never been satisfactorily determined. In most cases of such 'new' diseases planting material is moved to and fro on a global basis and the pathogens go with it, for example banana rhizomes were implicated in the spread of *F. oxysporum* f. sp. *cubense*.

Another possibility is that of a non-host carrying the pathogen. Price (1975c) found *F. oxysporum* f. sp. *narcissi* in a Norfolk field never previously cropped with *Narcissus* though there had been a crop of tulips in this field. Daffodils often follow tulips in crop rotations and it is possible that the tulip bulbs were infested elsewhere before they were brought to Norfolk.

It could be argued that vascular wilt of oil palm had been there earlier and had gone unnoticed or that it was an otherwise avirulent strain to which the hybrid palm was susceptible. This line of reasoning would also seem to apply to *F. oxysporum* f. sp. *tulipae* which caused so much damage when Darwin Hybrid tulips were introduced.

But there are other examples that have come to light in which the presence of *Fusarium* formae speciales defies explanation at the moment. Price and Linfield (unpublished) found 12 soil samples of 48 in a Cambridgeshire barley field contained *Fusarium oxysporum*. Of these 12, five contained *F. oxysporum* f. sp. *narcissi*. No bulbs had ever been planted in or near this field. The part played by root exudates, or appreciable amounts of soil moisture in temperate soils, is well documented but these factors would scarcely apply in deserts. With this in mind it is difficult to account for the presence of *F. avenaceum*, *F. equiseti*, *F. moniliforme*, *F. oxysporum*, *F. semitectum* and *F. solani* in desert-type soils (Joffe & Palti, 1977) all of which were alkaline (up to pH 9.0) and in regions with very hot day temperatures and little rain.

In this short section several examples of instances of *Fusarium* species occurring unexpectedly in soil have been quoted. Whilst they demonstrate ubiquity they also serve to show how little we really know about the ecology of pathogens of the genus.

Problems of symptomless carriers

The definition of a forma specialis is 'a taxon characterised from a physiological (especially host adoption) standpoint but scarcely or not at all from a morphological standpoint' (Federation of British Plant Pathologists, 1973). When this concept was devised and applied to *Fusarium* species narrow host ranges were implied or imagined. In some respects this is correct (or at this moment is correct) of a few formae speciales. For example *F. oxysporum* f. sp. *lycopersici* and *F. oxysporum* f. sp. *narcissi* seem to be confined to tomatoes and daffodils respectively. The position of others, once thought to have a limited host range, is now being questioned and re-examined. Thus, *F. oxysporum* f. sp. *gladioli* was thought to be restricted to *Gladiolus* and *Iris* but McClellan (1945) extended the list to include 11 other iridaceous genera.

Until a few years ago a forma specialis would have been identified by its characteristic damage to the plant it infested. As is so often the case, exceptions to a rule are soon found. The exceptions, in this case, are those that have come to be called 'symptomless' hosts. A few years earlier, before McClellan's work, symptomless hosts to *F. oxysporum* f. sp. *vasinfectum* had been found by Armstrong *et al.* (1941). They inoculated roots of sweet potato plants with the pathogen and although only 3% showed symptoms of wilt, the fungus was recovered from almost half of the plants.

Regrettably there is no strict definition of 'symptomless carrier'. This phrase and variants such as 'without symptoms' have been used for many years and an often accepted meaning has been a lack of external *visible* symptoms, e.g. wilting or necrosis on parts of the plant above ground. However, in the absence of proper definition these phrases are ambiguous. 'Symptomless carrying' does not necessarily mean that the disease organism is within a plant but can equally be present on its exterior. The two possible meanings are: first, a plant which is actively invaded but shows no visible symptoms such as Katan (1971) found with *Oryzopsis*, *Digitaria* and *Amaranthus* carrying *F. oxysporum* f. sp. *lycopersici*. Secondly, a plant or part of a non-host plant which carries the passive fungus externally. Two examples suffice: on seeds (Nash & Snyder, 1964) and *Narcissus* bulbs carrying *F. oxysporum* f. sp. *tulipae*. Narcissus crops in the United Kingdom frequently follow an annual crop of tulips and may carry the forma specialis *tulipae* on their fleshy scales. This latter example of a carrier can be further subdivided into two groups because *Fusarium* populations can increase on and in the roots of some plants but in others will slowly decline. Besri (1975) showed that wheat plants depressed populations of *F. oxysporum* whilst beans and tomatoes were associated with increased numbers.

These two concepts of a pathogen being carried within and on the outside of plants are relatively new in *Fusarium* investigations but the distinctions are not always made. Any conclusions drawn about which concept is correct will depend largely on the techniques used to isolate the pathogen from the symptomless host and then on the tests for pathogenicity. In isolations made from the bases of unsterilised root plates of *Narcissus* cv. Golden Harvest about half the isolates were pathogenic (Price, unpublished). Most workers surface-sterilise tissues from which *Fusarium* might be isolated but the techniques used vary considerably. It must be apparent that, whilst mycelium, macro- and microconidia are relatively easy to kill, chlamydospores are not by most sterilising techniques. If the sterilisation technique failed to kill externally borne chlamydospores the results might suggest that the fungus grew from within the tissues. Some chlamydospores of *F. oxysporum* f. sp. *narcissi* retained their viability after three hours in 0.5% formaldehyde (Price and Linfield, unpublished).

The literature on the topic of carriers is large but six examples will demonstrate the difficulties of interpretation. Armstrong & Armstrong (1948) developed a surface sterilisation method which showed that dried fungal deposits obtained from suspensions of cultures into which sticks had been dipped were killed by first dipping in 95% ethanol and then dipping for five minutes in a 5% calcium hypochlorite solution calculated to have 1667 p.p.m. available chlorine. Using this surface sterilising technique apparently healthy sweet potato roots were found to contain a *Fusarium* sp. Likewise both stem and roots of cotton plants were found to harbour *F. oxysporum* f. sp. *batatas*.

Hendrix & Nielsen (1958) dipped tap-water washed roots of sweet potato that had been inoculated with seven formae speciales, which included *F. oxysporum* f. sp. *batatas*, into a 1 : 10 dilution of a proprietary 5% sodium hypochlorite for ten minutes. Their results showed that *F. oxysporum* f. sp. *batatas* invaded and colonised roots and stems of 11 plant species. Of the other six formae speciales all invaded *Ipomoea* but only one caused symptoms.

MacDonald & Leach (1976) used the same technique when they discovered symptomless carriers of *F. oxysporum* f. sp. *betae* among common weeds in sugar beet. Although inoculated pigweed (*Amaranthus reflexas*) wilted, *Chenopodium album*, *Brassica nigra* and *Anethum graveolens* did not. Katan (1971) used a stronger solution, 1% sodium hypochlorite, for the shorter time of two minutes and found *F. oxysporum* f. sp. *lycopersici* within the tissues of *Oryzopsis*, *Digitaria* and *Amaranthus* species.

Kreutzer (1972) dipped roots of grassland plants for 30 seconds in 0.1% mercuric chloride solution. The principal *Fusarium* species were all chlamydospore-forming ones (*F. roseum*, *F. solani* and *F. oxysporum*). Haware & Nene (1982) concluded that *Cajanus cajan*, *Lens esculenta* and *Pisum seturium* harboured *F. oxysporum* f. sp. *ciceri*, the cause of chick pea wilt. Roots and stems were washed, then air dried before being cut into small pieces and sterilised for two minutes in 2.5% sodium hypochlorite.

In these six pieces of work the phrase 'symptomless host' or 'symptomless carrier' is either mentioned or implied. It is most important to distinguish between a symptomless host and a carrier. This must invariably mean histological evidence to prove or disprove tissue invasion.

A symptomless host must mean those plants within which *Fusarium* actively invades and grows extensively. Such plants should be included in the host range of a species or forma specialis. The term 'symptomless carrier' should be retained for those plants which support the active growth and actually contribute to an increase in the number of propagules without any fungal invasion of tissue.

Chlamydospores
Chlamydospore formation

The formation, survival and subsequent germination of chlamydospores are the likely keys to the carry-over of 18 species of plant pathogenic fusaria. In the broadest terms the cycle begins with the germination of a chlamydospore on or near a susceptible plant which probably leads to a transient increase in the soil population. Numbers decrease after the death of the host and the few remaining propagules are all chlamydospores. This may be an oversimplification for several reasons. That the carry-over is due to chlamydospores is probably correct in the long-term but there is evidence that macroconidia can retain their viability for a long time. *F. moniliforme* which does not produce chlamydospores can survive for several years in soil, presumably as macroconidia. Recently, Hargreaves & Fox (1977) have found thickened hyphae which seem to be survival structures. Plant residues, nutrient levels and 'non-susceptible' host plants may influence the size of this population. A review of this survival scheme was published in 1969 by Snyder.

It is over 60 years since Burkholder (1919) postulated that *F. solani* perennated as chlamydospores, a suggestion confirmed by Warcup (1955) and later, Nash, Christou & Snyder (1961). It is now possible to produce inocula of chlamydospores, to assay soil populations, and to assess disease. With these advances it is surprising that we are still unable to obtain better control of disease. One reason seems to be the contrasts provided by the published accounts of experimental results. There seems to be no single model of fungal behaviour.

In attempts to gain a greater understanding of chlamydospore formation, much effort has been put into *in vitro* studies. This work on chlamydospore formation generally falls into one of two categories. One is based on the premise that unfavourable conditions induce their formation; the other, that it is the fungal response to specific substances. Most *in vitro* studies have as their starting point a basal salt medium which allowed mycelial growth but not sporulation. Modifying this basal solution in a variety of ways has led to chlamydospore formation. Cochrane & Cochrane (1971) induced them to be formed in *F. solani* by using an acid salt medium. The low pH was achieved by using different sources and amounts of nitrogen. The growth medium was not deficient in carbon or nitrogen so chlamydospore formation was not caused by nutrient depletion. French & Nielsen (1966) believed chlamydospore formation was associated with a particular substance because they found them only in high concentrations of macroconidia and not in dilute suspensions. Barran, Schneider & Seaman (1977) also found this phenomenon.

However, the bulk of the *in vitro* work using defined substances has been done, not on specific inducers, but by manipulating carbon/nitrogen ratios or by using different sources of nitrogen. Thus Carlile (1956) recorded that a low carbon/nitrogen ratio favoured chlamydospore formation in *F. oxysporum*, while Griffin (1964) found, in contrast, that chlamydospore formation increased as the pH of the medium increased, providing glucose and potassium nitrate were both present. None were formed with glucose alone and only a few with potassium nitrate as the nitrogen source in the absence of glucose. Yet, Qureshi & Page (1970), using either *F. oxysporum* or *F. solani* (the text is ambiguous), found no effects from either ammonium or nitrate nitrogen. Increases in chlamydospore numbers occurred when carbon, both organic and inorganic, was added up to 2.0 mg/l. Above this, numbers declined.

Meyers & Cook (1972) induced synchronous chlamydospore formation when sucrose was suddenly removed from cultures of *F. solani* f. sp. *phaseoli* and *F. solani* f. sp. *pisi*. When the carbon source was allowed to decline slowly chlamydospore formation was slow. The results of Hsu & Lockwood (1973) also suggested that a declining carbon source induced chlamydospore formation in *F. oxysporum*, *F. solani* and *F. roseum*.

These *in vitro* experiments have shown clearly that chlamydospore formation is associated with levels of simple organic and inorganic chemicals. However, simple sugars, used in the laboratory, are unlikely to play a part in this process in soil. For example, Alexander *et al.* (1966) induced *F. solani* to form chlamydospores when sterile soil water extracts were added to cultures. Ford, Gold & Snyder (1970*a*, *b*) extended this work and found that there was a specific response by clones of *F. solani* f. sp. *phaseoli* to soil extracts and that the induction of chlamydospores differed seasonally. Three species of bacteria were found that caused this effect and seasonal fluctuation in chlamydospore numbers may be a response to other microbial populations. Bourret (1967) noted that temperature affected spore induction with soil extracts. Induction took place at 18–30 °C but not from 33–36 °C. Nyvall (1970) noted similar effects in unsterile soil with *F. roseum* 'Graminearum' where they were formed at 21–30 °C but not at 5–11 °C.

Prasad & Sinha (1972) observed more chlamydospores of *F. oxysporum* f. sp. *lathyri* in the hot dry season than the rainy season in India. It would seem, therefore, that there is a complex interaction of factors including soil temperature, soil moisture and the effects that these have on other microorganisms.

Kao & Lockwood (1976) found 30 isolates of *Streptomyces* that induced chlamydospore formation in *F. solani* f. sp. *phaseoli*. Several crop plants have been found to affect *Fusarium* populations possibly by affecting chlamydospore formation. Thus rape favoured organisms that are antagonistic to *F. oxysporum* f. spp. *lycopersici*, *phaseoli* and *F. nivale*. Cabbage roots caused the same effect on *F. oxysporum* but not *F. nivale*. *Penicillium* and *Trichoderma* seemed to be the most common antagonists (Lacicowa & Onlikowski, 1979).

Chlamydospore survival

Once formed, chlamydospores of formae speciales survive either in a dormant state or germinate to live saprophytically before forming chlamydospores once again, pending the arrival of a favoured host. A very prejudiced view this, where mycologists consider that some hosts (i.e. those that are damaged) are more favoured than others. Katan (1971) found *F. oxysporum* f. sp. *lycopersici* in the desert. Why should these isolates need this facility? How have they developed? We allocate importance to a forma specialis because it parasitises a plant and not because it exists saprophytically. However, this is a separate issue but it is clear that chlamydospores are able to survive for a long time in conditions unfavourable for mycelial growth. Booth (1971) gives some viabilities for 18 species surviving for three years in soil kept at room temperatures. It is particularly noteworthy that amongst this list are *F. avenaceum*, *F. moniliforme* and *F. poae*, three species that are not thought to form chlamydospores.

Compared with *in vitro* work the environment is very different in field soils which abound with many antagonistic organisms. The accuracy of measuring the ability of *Fusarium* to survive in soil is a function of technique. Whatever selective medium is used and whatever numbers of propagules/gram are quoted, much depends upon the pretreatment of the soil before dilution and the dilution used. One only has to compare the results of soil shaken and diluted in 0.1% water agar, 0.5% carboxymethyl cellulose or water to understand the magnitude of difference. Both numbers and uniformity of distribution are much improved with the more viscous diluents.

The other major problem is sample size and frequency of sampling. Many investigators pool soil samples and then by dilution calculate the overall size of the *Fusarium* population. If it is postulated that the pathogenic *Fusaria* are active on and in roots of a crop then the distribution will be related to the spatial arrangement of the host. This is particularly true in mono-cropping systems with extensive annual tillage operations. Thus, with short-lived closely spaced crops such as beans or cereals the distribution of the fungus will be more uniform than many other widely spaced perennial crops. In crops, e.g. bulbs, which are grown closely together in ridges, but with some distance between ridges, the pooling of random samples would be inadequate, since between ridges *Fusarium* would be scarce but within, plentiful. With crops planted relatively far apart such as bananas and oil palms, pooled or separate samples would be equally good.

There is a choice between recording numbers of *Fusarium* propagules in samples or numbers of samples containing *Fusarium*. Either way the amount of work, especially if pathogenicity tests are made, is daunting. Few people today can emulate Gordon's examination of 1674 soil samples yielding 12 485 isolates of *Fusarium* (Gordon, 1954), simply because of the time involved.

Quite apart from the amount of time needed in ecological studies there is a complete lack of a standard method. The diluent and its viscosity, type of agar and medium used for isolation will have a profound effect on numbers. This lack amongst ecologically minded *Fusarium* pathologists hinders comparative studies. Papavizas (1967*a*) used 17 different media but concluded that the simplicity of Nash medium gave an overwhelming advantage. He also (1967*b*) used carboxymethyl cellulose as a diluent to obtain an even spread of soil. Another pitfall in comparative work is the effect of agar on numbers. Marshall, Whiteside & Alexander (1960) compared numbers of micro-organisms in six soils using five media with two different agars. Space prevents going into the fine details of the techniques but the review by Tsao (1970) of selective media is particularly useful.

Once they have been formed, the behaviour of chlamydospores in the soil is governed by many factors. Because they represent the primary reservoir of infection any circumstance that leads to their germination has attracted much attention. Good disease control would follow if chlamydospores could be induced to germinate and lyse in the absence of a host plant. Research on this aspect has advanced in three areas: the effects of root exudates, organic matter amendments and physical conditions.

There have been two major reviews about plant exudate effects on microorganisms (Schroth & Hildebrand, 1964; Rovira, 1965) but since then several papers have been published which demonstrate the complexity of exudate effects. The nature of the exudates, chlamydospore age, seasonal effects (an integration of soil temperature and moisture) and soil nutrients as well as host and non-host plants all play a part. Probably because few of all these components can be studied at any one time some of the results seem to be in conflict.

Chlamydospore germination

Buxton (1962) compared the root exudates of two cultivated bananas, Gros Michel and Lacatan by chromatography. The former is susceptible to *F. oxysporum* f. sp. *cubense*, the latter resistant. Eighteen amino acids were found, of which 13 were common to both, and in addition there were eight common sugars. Exudates from Gros Michel stimulated chlamydospore germination whereas those from Lacatan did not, which suggests a specificity of effect. Kraft (1974) too found inhibitory effects with exudates from *F. oxysporum* f. sp. *phaseoli* resistant plants. On the other hand, Kommedahl (1966), after experimenting with races of *F. oxysporum* f. sp. *pisi*, concluded that exudates of three lines of peas triggered germination and that resistance was a quality associated with fungal penetration being prevented; Schippers & Voetburg (1969) confirmed these results. Ito & Ui (1975) found no differences in the effects of exudates from resistant and susceptible beans on germination of *F. solani* f. sp. *phaseoli* chlamydospores.

In the light of this evidence Buxton's results seem surprising but many factors other than exudates are involved in germination. Schroth & Snyder (1961) showed that chlamydospores of *F. oxysporum* f. sp. *pisi* only germinated near young roots, presumably because of a higher exudate concentration. Another important possibility which makes experimental inoculation techniques complex is that the age of the chlamydospores and the host plant may affect subsequent disease development. Schroth, Toussoun & Snyder (1963) found that, even under what were thought to be ideal conditions, not all chlamydospores of *F. solani* f. sp. *phaseoli* germinated. Age, leaching, fungistasis: there could be many explanations but, until the synchronous production of very large numbers of chlamydospores in a form suitable for inoculation is available, unambiguous experimental results will not be easy to achieve.

It is not surprising that the amount of soil moisture should have effects on chlamydospore germination if only because of the effect on concentration gradients of exudates. Cook & Flentje (1965) showed that germination and subsequent lysis of *F. solani* f. sp. *pisi* chlamydospores were indeed affected. Using soils with a range of moisture contents most spores germinated at 8.7%; four times as much disease was incited at this level than in dryer regimes.

Exudates and concentration gradients are, as with so much else, over-simplifications which do not take into account effects of other soluble salts. Elsewhere reference has been made to the effects of fertilisers, i.e. salts utilised by plants (and fungus?). Not all salts are of use to plants nutritionally and Besri (1981a, b) has shown that soil salinity itself has substantial effects on fungal populations and disease. *F. oxysporum* f. sp. *lycopersici* sporulates profusely in saline irrigation water but numbers fall rapidly as soil moisture decreases, i.e., as the salinity increases. Rintelen (1973a, b, c) approached the ecology of *Fusarium* differently by looking at the effects of host plants (peas) on populations of both *F. oxysporum* and *F. solani* growing with and without weeds. In weed-free soils virulent strains of *F. solani* increased whilst avirulent strains declined. Yet with *F. oxysporum* more virulent isolates were found in soil containing both peas and weeds.

Perhaps it is not the individual components of exudates that are important but the ratios of different amino acids and sugars. Such ratios would change if other organisms utilised them differentially. Claudius & Mehrotra (1973) seemingly found that effects of exudate or the exudates themselves differed with age. Working with *F. oxysporum* f. sp. *lentis*, they found ten amino acids and five sugars in the exudate of the lentil, *Lens culinaris*. Exudates from ten-day-old plants stimulated chlamydospore germination and were associated with more disease than those exudates from 21-day-old plants. Exudates from older plants contained the inhibitors glycine and phenylalanine. If this phenomenon is widespread it forms another pitfall for the unwary in which experimental results will differ according to time of inoculation.

Effects of organic amendments to soil

Many investigators have pursued the idea that amending the soil in some way might lead to a reduction in losses of yield caused by *Fusarium*. Oil seed cake (Singh & Singh, 1970), residues of cereals (Snyder, Schroth & Christou, 1959), chitin (Mitchell & Alexander, 1961a; Maurer & Baker, 1964), cellulose (Baker & Nash, 1965; Maurer & Baker, 1965) spent coffee beans (Adams, Lewis & Papavizas, 1968) and other crop residues (Sequeira, 1962) have all been used as amendments to *Fusarium* infested soil.

There is good evidence that chlamydospores are formed naturally as the nutrients available to the fungus diminish either as a result of leaching or because of competition from other organisms. The comparative saprophytic ability of *Fusarium* species differs considerably and some, e.g. *F. culmorum*, are strong competitors whilst *F. oxysporum* seems to be weak. Nonetheless, the pathogen's survival (as a pathogen) hinges upon a sufficiency of chlamydospores remaining viable until the arrival of a suitable host which not only induces germination, but also causes an increase in the number of propagules.

The effects of root exudates have been dealt with elsewhere. It is only a small step from the fortuitous natural event of a suitable root exudate causing germination to the deliberate addition of a substance to the soil that will have the same effect. If such an additive induces germination the chlamydospore loses the protection of the thick cell wall. Spores are induced to germinate and competitors and antagonists are also brought to activity which leads to lysis of the *Fusarium* mycelium. In laboratory studies Sequeira (1962) found that the average life of conidia of *F. oxysporum* f. sp. *cubense* in soil was about two weeks before lysis took place. Adding glucose stimulated chlamydospore germination and microconidium production but the spores soon lysed. Further chlamydospore formation was inhibited. However, chemicals such as simple sugars are expensive and for large scale operations crop residues are much cheaper; of velvet bean, cowpea and sugar cane residues Sequeira found the last to be most effective.

Patrick, Toussoun & Thorpe (1964) followed a similar line of investigation using *F. solani* f. sp. *phaseoli* chlamydospores. They found that cold water extracts of rye, barley, broccoli and broad bean plants stimulated germination. However, more lysis occurred with the rye and barley extracts. This work is interesting because it dovetails with some earlier work by Snyder, Schroth & Christou (1959), who found that crop rotations with barley reduced the subsequent amount of bean root rot, caused by *F. solani*, in the field. Experiments in glasshouses had shown that wheat straw, corn residues and pinewood shavings had this reducing effect whilst alfalfa and soybean residues increased the *Fusarium* population.

A high carbon to nitrogen ratio is known to suppress *F. solani* populations either because of competition from more successful organisms needing the nitrogen or because low nitrogen levels favour *Fusarium* antagonists. Baker & Nash (1965) added cellulose and ammonium nitrate separately or together. They found that the populations of *F. oxysporum* f. sp. *phaseoli* in the rhizosphere were unaffected, which supported the hypothesis of suppression through competition. However, Maurer & Baker (1965) used ammonium sulphate, glucose and cellulose to give different C/N ratios in naturally infested soils. A high N, low C ratio favoured disease, especially when ammonium rather than nitrate nitrogen was used.

Extending these lines of approach, Papavizas, Lewis & Adams (1968) added simple sugars and cellulose alone to chlamydospore-infested soils and recorded subsequent amounts of bean root rot. Several different soils were used, which also covered a range of pH 5.0 – 7.9. To these both ammonium and nitrate nitrogen were added to give different C/N ratios. Although analyses of results were complex the overriding result was that disease incidence was related to amounts of nitrogen.

Veldkamp (1955) showed that chitin stimulated the growth of actinomycetes: known antagonists of *Fusarium*. Mitchell & Alexander (1961*a*, *b*) and Maurer & Baker (1964) also used chitin but the results were obscure. Mitchell & Alexander found that of several additives only chitin appreciably reduced disease caused by *F. solani* f. sp. *phaseoli*. The same authors noted that when chitin was added to soil the fungal population was much reduced although the proportion of *Penicillium* colonies and numbers of actinomycetes also increased. This effect lasted for about two weeks. However, when Maurer & Baker (1964) added either lignin or chitin no difference in disease incidence was seen, although when added together there was less infection. Addition of potassium nitrate led to further reduction in the amounts of disease.

An unusual additive is that of spent coffee beans from instant coffee manufacture (Adams, Lewis & Papavizas, 1968). As with chitin, the effect occurred within a comparatively short time. Chlamydospore germination of *F. solani* was induced within about eight hours of the residue being added. Germ-tube lysis was seen shortly afterwards and the population was halved between seven and 14 days later. The fungistatic effect on chlamydospores did not occur at this time but only came into being from 4–28 days afterwards.

Toussoun, Menzinger & Smith (1969) used conifer litter as a stimulant for chlamydospore germination and subsequent germ-tube lysis. Lindeman (1970) found many *Fusarium* propagules in a meadow soil near a pine forest. The forest soil contained few propagules and water extracts of this soil caused a high germination followed by lysis of *F. oxysporum* chlamydospores from the meadow soil.

Effect of inorganic additives to soil

The development and severity of most diseases are affected by the nutrient status of the soil. Experiences and observations of farmers have often led to controlled experiments to study these aspects in detail. The converse is also true and there are many accounts of fertiliser experiments in which effects on disease have been discovered. Garrett (1956) listed the effects of nitrogen and potash fertilisers on eight formae speciales of *F. oxysporum*. Walker & Hooker (1945) with cabbage wilt, Walker & Foster (1946) with tomato wilt, McClellan & Stuart (1947) with narcissus basal rot and Stoddard (1947) with musk melon wilt found that high nitrogen coupled with low potash favoured disease. Others have recorded the single effects of nitrogen or potash.

Wensley & McKeen (1964) examined the effects of fertilisers in detail on both soil populations and disease incidence of *F. oxysporum* f. sp.

melonis. Artificially and naturally infested soils were used to which two amounts of conidia and three rates of 10.10.10 NPK fertiliser were added. The larger the amount of fertiliser the sooner and more severe the degree of wilt. Other experiments which involved calcium, potassium and ammonium nitrate additions were also made. The general conclusion was that musk melon wilt develops faster with a good nutrient supply than with a poor one, and confirmed the work of Stoddard (1947) that available nitrogen is of prime importance to the disease. The results with different populations of *Fusarium* suggested that even these are governed by the nutrient status of the soil. A small population with good nutrients is as potentially dangerous as a large one in a poor soil.

All these results, which are consistent, were obtained with fertilisers applied to the soil. Curiously, Bloom & Walker (1955) recorded less disease on tomato if urea was applied as a foliar spray before inoculation and more when it was applied afterwards.

Constraints on disease development

Most of the diseases caused by *Fusarium* have proved to be very difficult to control. Physical, chemical, biological and genetic methods have all been used with differing degrees of success. In almost all cases good crop husbandry plays a major part in denying any advantage to the pathogen, usually by preventing parasitic activity of the chlamydospores.

Disease escape is always worth seeking because the cost of this method is usually slight. A degree of escape can be achieved either by adequate crop rotations or by planting the host at distances too great for *Fusarium* to move from one plant to another within the lifetime of the host. The latter method may be satisfactory for an annual crop but may fail with biennial or perennial ones. *Narcissus* is usually grown as a biennial crop in the United Kingdom and basal rot, caused by *F. oxysporum* f. sp. *narcissi*, is the biggest single cause of yield loss. When bulbs were planted 7 cm distant from infected bulbs no infection took place after one season whilst it was considerable in the second (Price, 1977). Even with crop plants such as oil palms as widely spaced as 10 m, *Fusarium* spreads rapidly. Although in the wet tropics this might be, in part, caused by water-borne inocula it is more likely to be the result of roots of nearby healthy plants growing into the space freed from competition, and becoming infested by the dying plant.

Soil temperature may play a part in disease escape. For example, Burke (1965) and Burke & Nelson (1965) showed that wide spacing lessened the incidence of bean root rot, caused by *F. solani*, at warm temperatures. At lower temperatures, about 16 °C, this effect of distance on disease incidence was negated, presumably because the host grew less vigorously. However, the interaction between soil temperature and plant growth relative to fungal activity is complex. Price (1981) showed that *F. oxysporum* f. sp. *narcissi* is active all the year round in the ridges in which bulbs are grown in England. It was also shown (Price, 1977) that narcissus roots begin to senesce in May as the soil temperatures rise so that both host tissue and temperature favour the fungus. Lifting bulbs earlier than usual, i.e. in June instead of late July, gave a degree of control. Time in this case became another form of spatial escape, with time of lifting and annual instead of biennial cropping being important factors in disease control.

Bergman (1965) and Bergman & Noordermeer-Luyk (1973) showed that, in a similar disease of tulips, temperature – though important – was not the decisive factor affecting time of infection. In the early part of the year water soluble compounds inhibiting growth by *F. oxysporum* f. sp. *tulipae* are present in the outermost fleshy scale of the tulip bulb. At maturity, this outermost scale is membraneous and forms the brown tunic. As maturity is reached the compounds decline with a corresponding increase in disease susceptiblity. The effect of increasing soil temperature with advancing season was correlated with disease incidence by Gould & Miller (1975) in the United States but, as they did not seek fungitoxic substances, the two factors were unreported.

Soil temperature effects have been noted by several other investigators. Reyes (1970) using a range of temperatures from 22–26 °C found that the temperature at and immediately after inoculation was critical for disease development. Cabbage plants inoculated with *F. oxysporum* f. sp. *conglutinans* which were grown at 26 °C for a period and then at 22 °C developed more severe symptoms than those grown at 22 °C and then 26 °C.

F. oxysporum f. sp. *pisi* has an optimum temperature for growth equal to that of pea plants at about 25 °C. Maraite & Mayer (1967) failed to record any symptom in inoculated susceptible pea plants grown at 15 °C even though both host and pathogen grew satisfactorily at the sub-optimum level. An early record of temperature effects on symptom expression was that of Clayton (1923). Infected tomatoes wilted at temperatures between 25 and 30 °C with a slow blight (to quote his words) in a band on either of these temperatures but with no visible symptoms developing below 20 °C or above 30 °C.

Amongst the reports of physical effects on *Fusarium* those in which carbon dioxide have been used are interesting. High concentrations have been shown to activate (mycelial) infections in bulbs (Magie, 1971). This advanced development of disease gives the opportunity of reducing numbers of diseased bulbs before planting by sorting healthy bulbs and corms from diseased ones visually. Snyder (1968) found that mycelial growth was stimulated and chlamydospore formation inhibited in both *F. solani* and *F. oxysporum*. Bourret, Gold & Snyder (1968) enlarged on this discovery, finding that carbon dioxide stimulated chlamydospore germination in soil only when organic nutrients were available. These results taken together suggest that the use of carbon dioxide coupled with fungicides to free dormant plants from *Fusarium* infestation is possible.

There is good evidence that with the principal *Fusarium* pathogens it is the chlamydospores that provide the seasonal carry-over and that these form the primary reservoir of infection. The secondary reservoirs of infection are conidia and mycelial fragments on seed and other parts of plants. These may be relatively easily controlled by direct chemical means. Chlamydospores, which are able to remain viable though dormant for many years, are less easily attacked in this way. Indirect methods such as crop rotations and soil amendments seem likely to be the most successful.

It would seem that the next step forward in combating the huge losses of food caused by *Fusarium* is for pathologists to standardise some of the methods used for isolating the fungus. Once this happens a much more general understanding of the behaviour of the pathogen away from favoured hosts will be achieved. Understanding the size of the primary reservoir of infection and manipulating it to reduce it will lead to substantially less disease. No mention has been made of the potential problems caused by heterokaryosis or parasexuality or of breeeding for resistance. There have been some notable successes in this, but that story belongs elsewhere.

References

Adams, P. B., Lewis, J. A. & Papavizas, G. C. (1968). Survival of root-infecting fungi in soil. 9. Mechanism of control of root rot of bean with spent coffee beans. *Phytopathology*, **58**, 1603–8.

Alexander, J. V., Bourret, J. A., Gold , A. H. & Snyder, W. C. (1966). Induction of chlamydospore formation by *Fusarium solani* in sterile soil extracts. *Phytopathology*, **56**, 353–5.

Armstrong, G. M. & Armstrong, J. K. (1948). Non-susceptible hosts as carriers of wilt fusaria. *Phytopathology*, **38**, 808–26.

Armstrong, G. M., Shanor, L., Bennett, C. C. & Hawkins, B. S. (1941). Some *Fusarium* wilt organisms. *Phytopathology*, **31**, 1–2.

Baker, R. & Nash, S. M. (1965). Ecology of plant pathogens in soil. 6. Inoculum density of *Fusarium solani* f. sp. *phaseoli* in bean rhizosphere as affected by cellulose and supplemental nitrogen. *Phytopathology*, **55**, 1381–2.

Barran, L. R., Schneider, E. F. & Seaman, N. L. (1977). Requirements for the rapid conversion of macroconidia of *Fusarium sulphureum* to chlamydospores. *Canadian Journal of Microbiology*, **23**, 148–51.

Bergman, B. H. H. (1965). Field infection of tulip bulbs by *Fusarium oxysporum*. *Netherlands Journal of Plant Pathology*, **71**, 129–35.

Bergman, B. H. H. & Noordermeer-Luyk, C. E. I. (1973). Influence of temperature on field infection of tulip bulbs by *Fusarium oxysporum*. *Netherlands Journal of Plant Pathology*, **79**, 221–8.

Besri, M. (1975). [Research on *Fusarium* diseases. Effect of crop predecessor on the evolution of the *Fusarium oxysporum* population in the rhizosphere of some plants.] *Annales Phytopathologie*, **7**, 1–8.

Besri, M. (1981*a*). Influence de la salinité du sol et des eaux d'irrigation sur la population de *Fusarium oxysporum* (Schl.) f. sp. *lycopersicae* (Sacc.) Snyd, et Hans. *Phytopathologie Mediterranean*, **20**, 101–6.

Besri, M. (1981*b*). Qualité des sols et des eaux d'irrigation et manifestation des trachéomycoses de la Tomate au Maroc. *Phytopathologie Mediterranean*, **20**, 107–11.

Bloom, J. F. & Walker, J. C. (1955). Effect of nutrient sprays on *Fusarium* wilt of tomato. *Phytopathology*, **45**, 443–4.

Booth, C. (1971). *The genus* Fusarium. Commonwealth Mycological Institute, Kew, Surrey.

Bourret, J. A. (1967). Physiology of chlamydospore formation and survival of *Fusarium*. *Dissertation Abstracts*, **27**, 2223.

Bourret, J. A., Gold, A. H. & Snyder, W. C. (1968). Effect of carbon dioxide on germination of chlamydospores of *Fusarium solani* f. sp. *phaseoli*. *Phytopathology*, **58**, 710–11.

Burke, D. W. (1965). Plant spacing and *Fusarium* rot of beans. *Phytopathology*, **55**, 757–9.

Burke, D. W. & Nelson, C. E. (1965). Effects of row and plant spacings on yields of dry beans in *Fusarium* infested and non-infested fields. *Washington Agricultural Experiment Station Bulletin* 664.

Burkholder (1919). The dry root-rot of the bean. *Cornell University Agricultural Research Station Memoir*, **26**, 999–1033.

Buxton, E. W. (1962). Root exudates from banana and their relationship to strains of *Fusarium* causing Panama wilt. *Annals of Applied Biology*, **50**, 269–82.

Carlile, M. J. (1956). A study of the factors influencing non-genetic variation in a strain of *Fusarium oxysporum*. *Journal of General Microbiology*, **14**, 643–54.

Claudius, G. R. & Mehrotra, R. S. (1973). Root exudates from lentil (*Lens culinaris medic*) seedlings in relation to wilt disease. *Plant and Soil*, **38**, 315–20.

Clayton, E. E. (1923). The relationship of temperature to the *Fusarium* wilt of the tomato. *American Journal of Botany*, **10**, 71–88.

Cochrane, V. W. & Cochrane, Jean C. (1971). Chlamydospore induction in pure culture in *Fusarium solani*. *Mycologia*, **63**, 462–79.

Cook, R. J. & Flentje, N. T. (1965). Influence of soil water on growth and survival of *Fusarium solani* f. sp. *pisi* in the rhizosphere of pea. *Phytopathology*, **55**, 1051–85.

De Munk, W. J. (1971). Bud necrosis in tulip bulbs, a multifactorial disorder (temperature, ethylene, mites). *Acta Horticulturae*, **23**, 242–8.

Federation of British Plant Pathologists (1973). *A Guide to the Use of Terms in Plant Pathology.* Commonwealth Mycological Institute, Kew, Surrey.

Ford, E. J., Gold, A. H. & Snyder, W. C. (1970a). Soil substances inducing chlamydospore formation by *Fusarium*. *Phytopathology*, **60**, 124–8.

Ford, E. J., Gold, A. H. & Snyder, W. C. (1970b). Induction of chlamydospore formation in *Fusarium solani* by soil bacteria. *Phytopathology*, **60**, 479–84.

French, E. R. & Nielsen, L. W. (1966). Production of macroconidia of *Fusarium oxysporum* f. sp. *batatas* and their conversion to chlamydospores. *Phytopathology*, **56**, 1322–3.

Garrett, S. D. (1956). *Biology of Root-Infecting Fungi.* Cambridge University Press.

Gordon, W. L. (1954). The occurrence of *Fusarium* species in Canada. 4. Taxonomy and prevalence of *Fusarium* species in the soil of cereal plots. *Canadian Journal of Botany*, **32**, 622–9.

Gould, C. J. & Miller, V. L. (1975). Effect of time of digging on incidence of *Fusarium* rot in tulip bulbs. *Acta Horticulturae*, **47**, 119–23.

Griffin, G. J. (1964). Influence of carbon and nitrogen nutrition on chlamydospore formation by *Fusarium solani* f. sp. *radicicola*. *Phytopathology*, **54**, 894.

Hargreaves, A. J. & Fox, R. A. (1977). Survival of *Fusarium avenaceum* in soil. *Transactions of British Mycological Society*, **69**, 425–8.

Haware, M. P. & Nene, Y. L. (1982). Physiologic races of the chickpea wilt pathogen. Symptomless carriers of the chickpea wilt fungus. *International Chickpea Newsletter*, **1**, 7–8.

Hendrix, F. F. & Nielsen, L. W. (1958). Invasion and infection of crops other than the forma concept by *Fusarium oxysporum* f. *batatas* and other formae. *Phytopathology*, **48**, 224–8.

Hsu, S. C. & Lockwood, J. L. (1973). Chlamydospore formation in *Fusarium* in sterile salt solutions. *Phytopathology*, **63**, 597–602.

Ito, I. & Ui, T. (1975). [Behaviour of *Fusarium* f. sp. *solani* in the rhizosphere and underground tissues of susceptible and non-susceptible plants.] *Memoirs of the Faculty of Agriculture, Hokkaido University*, **9**, 187–92.

Joffe, A. Z. & Palti, J. (1977). Species of *Fusarium* found in uncultivated desert type soils in Israel. *Phytoparasitica*, **5**, 119–21.

Kao, M. I. & Lockwood, J. L. (1976). Induction of chlamydospore formation in *Fusarium* on agar media. *Summa Phytopathologica*, **2**, 292–8.

Katan, J. (1971). Symptomless carriers of the tomato *Fusarium* wilt pathogen. *Phytopathology*, **61**, 1213–37.

Kommedahl, T. (1966). Relation of exudates of pea roots to germination of spores in races of *Fusarium oxysporum* f. sp. *pisi*. *Phytopathology*, **56**, 721–2.

Kraft, J. M. (1974). The influence of seedling exudates on the resistance of peas to *Fusarium* and *Pythium* root rot. *Phytopathology*, **64**, 190-3.

Kreutzer, W. A. (1972). *Fusarium* species as colonists and potential pathogens in root zones of grassland plants. *Phytopathology*, **62**, 1066–70.

Lacicowa, B. & Onlikowski, L. (1979). Studies on the influence of some cultivated plants upon mycoflora of the soil environment in phytopathological aspect. *Roczniki Nauk Rolniczych*, **6**, 7–35.

Lindeman, R. G. (1970). Plant residue decomposition products and their effects on host roots and fungi pathogenic to roots. *Phytopathology*, **60**, 19–21.

McClellan, W. D. (1945). Pathogenicity of the vascular *Fusarium* of gladiolus to some additional iridaceous plants. *Phytopathology*, **35**, 921–30.

McClellan, W. D. & Stuart, N. W. (1947). Nutrition and *Fusarium* basal rot in narcissus. *American Journal of Botany*, **34**, 88–93.

MacDonald, J. D. & Leach, L. D. (1976). Evidence for an extended host range of *Fusarium oxysporum* f. sp. *betae*. *Phytopathology*, **66**, 822–7.

Magie, R. O. (1971). Carbon dioxide treatment of gladiolus corms reveals latent *Fusarium* infections. *Plant Disease Reporter*, **55**, 340–1.

Maraite, H. & Mayer, J. A. (1967). Research on *Fusarium* diseases. 4. Effect of temperature on fusaria wilt of peas. *Annales Epiphytique* (Paris), **18**, 305–12.

Marshall, K. C., Whiteside, Jean S. & Alexander, M. (1960). Problems in the use of agar for the enumeration of soil micro-organisms. *Proceedings of the American Society for Soil Science*, **24**, 61–2.

Maurer, C. L. & Baker, R. (1964). Ecology of plant pathogens in soil. 1. Influence of chitin and lignin amendments on development of bean root rot. *Phytopathology*, **54**, 1425–6.

Maurer, C. L. & Baker, R. (1965). Ecology of plant pathogens in soil. 2. Influence of glucose, cellulose and inorganic nitrogen amendments on development of bean root rot. *Phytopathology*, **55**, 67–72.

Meyers, J. A. & Cook, R. J. (1972). Induction of chlamydospore formation in *Fusarium solani* by abrupt removal of the organic carbon substrate. *Phytopathology*, **62**, 1148–53.

Mitchell, R. & Alexander, M. (1961a). The mycolytic phenomenon and biological control of *Fusarium* in the soil. *Nature, London*, **190**, 109–10.

Mitchell, R. & Alexander, M. (1961b). Chitin and the biological control of *Fusarium* diseases. *Plant Disease Reporter*, **45**, 487–90.

Nash, S. M., Christou, T. & Snyder, W. C. (1961). Existence of *Fusarium solani* f. sp. *phaseoli* in soil. *Phytopathology*, **51**, 308.

Nash, S. M. & Snyder, W. C. (1962). Quantitative estimations of plate counts of propagules of the bean root rot *Fusarium* in field soils. *Phytopathology*, **52**, 567.

Nash, S. M. & Snyder, W. C. (1964). Dissemination of the root rot *Fusarium* with bean seed. *Phytopathology*, **54**, 880.

Nyvall, R. F. (1970). Chlamydospores of *Fusarium roseum* 'graminearum' as survival structures. *Phytopathology*, **60**, 1175–7.

Papavizas, G. C. (1967a). Evaluation of various media and antimicrobial agents for isolation of *Fusarium* from soil. *Phytopathology*, **57**, 848–52.

Papavizas, G. C. (1967b). Survival of root infecting fungi in soil. A quantitative propagule assay method of observation. *Phytopathology*, **57**, 1242–6.

Papavizas, G. C., Lewis J. A. & Adams, P. B. (1968). Survival of root infecting fungi in soil. 2. Influence of amendment and soil carbon : nitrogen balance on *Fusarium* root rot of beans. *Phytopathology*, **58**, 365–72.

Patrick, Z. A., Toussoun, T. A. & Thorpe, H. J. (1964). Influence of Plant-derived toxic decomposition products on development of *Fusarium solani* f. *phaseoli* and *Thielaviopsis basicola*. *Proceedings of Canadian Phytopathological Society*, 31.

Prasad, T. & Sinha, S. (1972). Studies on the survival of *Fusarium oxysporum* f. sp. *lathryi* causing wilt of khesari (*Lathryrus sativus*) in soil. *Indian Phytopathology*, **25**, 423–7.

Price, D. (1975a). Pathogenicity of *Fusarium oxysporum* found on narcissus bulbs and in soil. *Transactions of the British Mycological Society*, **64**, 137–42.

Price, D. (1975b). Pathogenesis of tulips by *Fusarium oxysporum*. *Transactions of the British Mycological Society*, **64**, 550–2.

Price, D. (1975c). The occurrence of *Fusarium oxysporum* in soils, and on narcissus and tulip. *Acta Horticulturae*, **47**, 113–18.

Price, D. (1977). Some pathological aspects of narcissus basal rot, caused by *Fusarium oxysporum* f. sp. *narcissi*. *Annals of Applied Biology*, **86**, 11–17.

Price, D. (1981). Biology of bulb pathogens. Basal rot of narcissus caused by *Fusarium oxysporum* f. sp. *narcissi*. *Glasshouse Crops Research Institute Report 1980*, pp. 131–2.

Qureshi, A. A. & Page, O. T. (1970). Observations on chlamydospore production by *Fusarium* in a two salt solution. *Canadian Journal of Microbiology*, **16**, 29–32.

Reyes, A. A. (1970). The effect of soil temperature on *Fusarium* yellows of cabbage. *Proceedings of the Canadian Phytopathological Society*, **37**, 28.

Rintelen, J. (1973a). [Influence of weeds on the infection of peas and flax by soil borne fusaria. 1. Observations on a long term experiment and investigations on populations of pathogens.] *Zeitschrift für Pflanzenkrankheiten und Pflanzenschutz*, **80**, 265–83.

Rintelen, J. (1973b). [Influence of weeds on the infection of peas and flax by soil borne fusaria. 2. Investigations into the influence of the rhizosphere on the composition of the fusaria populations in the soil.] *Zeitschrift für Pflanzenkrankheiten und Pflanzenschutz*, **80**, 395–402.

Rintelen, J. (1973c). [Influence of weeds on the infection of peas and flax by soil borne fusaria. 3. Experiments on competition between pathogenic and apathogenic isolates of *Fusarium solani* and *Fusarium oxysporum* in infection of peas.] *Zeitschrift für Pflanzenkrankheiten und Pflanzenschutz*, **80**, 466–70.

Rovira, A. D. (1965). Plant root exudates and their influence upon semi-microorganisms. In *Ecology of Soil Borne Pathogens*, ed. K. R. Baker & W. C. Snyder, pp. 170–4. London: John Murray.

Schippers, B. & Voetburg, J. (1969). Germination of chlamydospores of *Fusarium oxysporum* f. sp. *pisi* race 1 in the rhizosphere, and penetration of the pathogen into roots of a susceptible and resistant pea cultivar. *Netherlands Journal of Plant Pathology*, **74**, 241–58.

Schroth, M. N. & Hildebrand, D. C. (1964). Influence of plant exudates on root infecting fungi. *Annual Review of Phytopathology*, **2**, 101–32.

Schroth, M. N. & Snyder, W. C. (1961). Effect of host exudates on chlamydospore germination of the bean root rot fungus, *Fusarium solani* f. sp. *phaseoli*. *Phytopathology*, **51**, 390–3.

Schroth, M. N., Toussoun, T. A. & Snyder, W. C. (1963). Effect of certain constituents of bean exudate on germination of chlamydospores of *Fusarium solani* f. sp. *phaseoli*. *Phytopathology*, **53**, 809–12.

Sequeira, L. (1962). Influence of organic amendments on survival of *Fusarium oxysporum* f. sp. *cubense* in soil. *Phytopathology*, **52**, 976–82.

Singh, R. S. & Singh, N. (1970). Effect of oil-cake amendment of soil on populations of some wilt causing species of *Fusarium*. *Phytopathologische Zeitschrift*, **69**, 160–7.

Snyder, W. C. (1968). The effect of carbon dioxide on the germination of chlamydospores of *Fusarium solani*. *Phytopathology*, **58**, 710–11.

Snyder, W. C. (1969). Survival of *Fusarium* in soil. *Annales de Phytopathologie*, **1**, 209–12.

Snyder, W. C., Schroth, M. N. & Christou, T. (1959). Effect of plant residues on root rot of bean. *Phytopathology*, **49**, 755–6.

Stoddard, D. L. (1947). Nitrogen, potassium and calcium in relation to *Fusarium* wilt of musk melon. *Phytopathology*, **37**, 875–84.

Toussoun, T. A., Menzinger, W. & Smith, R. S. (1969). Role of conifer litter in ecology of *Fusarium* : stimulation of germination in soil. *Phytopathology*, **59**, 1396–9.

Tsao, P. H. (1970). Selective media for isolation of pathogenic fungi. *Annual Review of Phytopathology*, **8**, 157–86.

Veldkamp, H. (1955). Aerobic decomposition of chitin by microorganisms. *Mededelingen van de Landbouwhoogeschoolte Wageningen*, **55**, 127–74.

Walker, J. C. & Foster, R. E. (1946). Plant nutrition in relation to disease development. III. Tomato Wilt. *American Journal of Botany*, **33**, 259–64.

Walker, J. C. & Hooker, W. J. (1945). Plant nutrition in relation to disease development. I. Cabbage Yellows. *American Journal of Botany*, **32**, 314–20.

Warcup, J. H. (1955). On the origin of colonies of fungi developing on soil dilution plates. *Transactions of the British Mycological Society*, **38**, 298–301.

Wensley, R. N. & McKeen, C. D. (1964). Some relationships between plant nutrition, fungal populations, and incidence of *Fusarium* wilt of muskmelon. *Canadian Journal of Microbiology*, **11**, 581–94.

5

The production and significance in phytopathology of toxins produced by species of *Fusarium*

Department of Microbiology, The University of Birmingham, P.O. Box 363, Birmingham B15 2TT, U.K.

Introduction

Many species of *Fusarium*, in common with at least some members of each of the major taxonomic groups of fungi pathogenic to plants (Rudolph, 1976), produce, in culture, compounds which are toxic to plants. The precise role of such toxins in the development of disease has usually not been established but for most of them it is probably mainly a role in the extension of the lesion rather than in the determination of host specificity.

Toxins produced by pathogens of plants may be classified in a number of ways (Yoder, 1980; Scheffer & Briggs, 1981) which take into account the properties of the toxins and their function in the development of disease. With regard to the latter, toxins are frequently classified as host-specific (host-selective might be better) or as non-host-specific. Host-specific toxins are highly active against the host of the pathogen producing the toxins, and these hosts are often cultivars of a single species. Such toxins have little or no permanent effect, even at concentrations several orders of magnitude higher than those affecting sensitive plants, on cultivars resistant to the pathogen or on non-host species. The extreme specificity, shared with the respective pathogen, of host-specific toxins for plants of susceptible cultivars suggests a role for such toxins in both disease development and incitement.

All known toxins, except possibly phytonivein (Hiroe & Nishimura, 1956), produced by species of *Fusarium* belong to the second category, i.e. non-host-specific toxins. These produce only some of the symptoms of the disease and may affect a wide range of plants not solely the hosts of the fungal species producing the toxins – i.e. the range of plants sensitive to the toxin is greater than the host range of the fungus. A number of these toxins also affect animals (see chapters 7 and 9).

When grown in culture fungi produce numerous compounds which are toxic to plants and to establish that a compound has a causal role in disease development it should be examined using as many criteria as possible. These include:

(a) Isolation of the toxin from, or identification of the toxin in, diseased plants and the demonstration that it is present in a suitable concentration to cause the symptoms attributed to it.

(b) Production of typical disease symptoms when isolated toxin is applied to healthy plants at concentrations similar to those found in infected plants.

(c) Correlation of the virulence of isolates of the fungus with their ability to produce toxin. This is usually determined using *in vitro* cultures although measurement of toxin production *in vivo* is desirable.

However, it may not be possible to satisfy these criteria; nor are they entirely satisfactory (Durbin & Steele, 1979; Yoder, 1980; Scheffer & Briggs, 1981) for a number of reasons. For example, it may not be possible to isolate the toxin because it is too labile, present in a concentration too small to be measured, is tightly bound to host tissue or is inactivated by the host. Secondly, since plants are limited in the ways in which they can respond to injury, the production of visible or physiological symptoms by application of a toxin does not prove that the toxin causes the symptoms when the plant is infected by a fungus capable of producing the toxin *in vitro*. Conversely, since the range of symptoms produced in a disease probably has a multiplicity of causal agents, it may be unreasonable to expect an isolated, purified toxin to produce all of the symptoms typical of the disease. Thirdly, toxin production *in vitro* may be markedly influenced by cultural conditions and different isolates of a fungus may not respond in the same way to changes in those conditions (Egli, 1969; Yoder, 1980). Thus virulence may not correlate with the quantity of toxin produced *in vitro*. Toxin production *in vivo* would be a better criterion but, even when feasible, this is usually not determined.

From these reservations it should be evident that even if a toxin conforms to all the criteria it does not necessarily have a causal role in pathogenesis nor is a compound which does not entirely conform excluded from being a toxin with such a role.

Perhaps the best test to establish that a toxin has a causal role in the development of a disease is a comparison of the progress of the disease in the presence of and following .the specific elimination of the toxin. Elimination can be achieved in a number of ways: the best of these is deletion (either induced or natural) of all or part of a structural gene involved in the production of the toxin, since this should leave the remainder of the system unchanged with virtually no chance of reversion to toxin production.

At least 20 toxins have been described from about 30 species of *Fusarium* (Table 5.1). Structures have been proposed for most of these toxins but the suggested structure has been confirmed by synthesis for less than half of them. The toxins have widely differing chemical constitutions and may differ considerably in their mode of action and in the symptoms elicited in susceptible plants.

Several of the toxins are considered from the viewpoint of structure, mode of action and role in the host–parasite relationship. For the other toxins, while the structures are known there is little information available relating to mode of action or role in pathogenesis.

Lycomarasmin

Lycomarasmin (Fig. 5.1), which is produced by a number of formae speciales of *F. oxysporum*, was probably the first compound toxic to plants to be isolated in pure form from cultures of a plant pathogen (Clauson-Kaas, Plattner & Gäumann, 1944). However, it was nearly 20 years later before its structure was finally established (Hardegger *et al.*, 1963).

Treatment of tomato cuttings with 10^{-2} M lycomarasmin (Gäumann, 1951) causes first a decrease in water uptake, and in transpiration, and then an increase in both such that transpiration exceeds uptake and desiccation occurs – Gäumann's 'Pathological Wilting'. The high concentrations of lycomarasmin required to damage healthy tomato cuttings cause an upward rolling of the leaflets, interveinal necrosis and desiccation of the leaf lamina without wilting.

In common with many *Fusarium* toxins lycomarasmin chelates metal ions. It forms an unstable chelate with iron which is much more toxic than is free lycomarasmin. The chelate is translocated to the leaves where the iron is released causing necrosis of the leaf tips. The chelate also causes a rapid increase in respiration of the leaves.

Lycomarasmin has not been detected in inoculated plants although it is claimed (Kern, 1972) that it is produced in a few days, i.e. at an early stage of growth, in culture. Other work (Dimond & Waggoner, 1953), however, suggests that lycomarasmin, in common with most toxins

Table 5.1. *Toxins produced by species of* Fusarium

Species	Toxin	Reference
F. avenaceum	Enniatins A & B	Braun, 1963; Ballio, 1972
F. batatis	Fusaric acid	Nishimura, 1957; Singh
	Dehydrofusaric acid	& Husain, 1970; Kern, 1972
F. conglutinans	Fusaric acid	See *F. batatis*
	Dehydrofusaric acid	
F. cubense	Fusaric acid	See *F. batatis*
	Dehydrofusaric acid	
F. culmorum	Culmonomarasmin	Kiss et al., 1960
F. equiseti	12,13-epoxytrichothecenes	Brian et al., 1961
F. fusarioides	Moniliformin	Rabie et al., 1978
F. gibbosum	Enniatins	Minasyan et al., 1978
F. heterosporum	Fusaric acid	See *F. batatis*
	Dehydrosfusaric acid	
F. javanicum	Naphthazarins	Kern & Naef-Roth, 1966; Dorn, 1974; Kern, 1978
F. lateritium	Enniatins	Bishop & Ilsley, 1978;
	α-picolinic acid	Berestetskii et al., 1976
F. lini	Fusaric acid	See *F. batatis*
	Dehydrofusaric acid	
F. lycopersici	Fusaric acid	See *F. batatis*
	Dehydrofusaric acid	
	10-hydroxyfusaric acid	Ruffner, 1974
F. martii	Naphthazarins	See *F. javanicum*
F. niveum	Fusaric acid	See *F. batatis*
	Dehydrofusaric acid	
F. orthoceras	Enniatins A & B	See *F. avenaceum*
	Fusaric acid	*See F. batatis*
	Dehydrofusaric acid	
F. oxysporum	Enniatins A & B	See *F. avenaceum*
F. oxysporum f. sp.		
auranthicum	Enniatins A & B	See *F. avenaceum*
lycopersici	Fusaric acid	Kuo & Scheffer, 1964
lycopersici	Lycomarasmin	Kern, 1972
	Lycomarasmic acid (Aspergillomarasmin B)	Camporota et al., 1973
	Aspergillomarasmin A	
	Phytolycopersin	Kern, 1972
melonis	Lycomarasmin	See *F. o.* f. sp.
	Lycomarasmic acid	*lycopersici*
	Aspergillomarasmin B	
niveum	Phytonivein	Hiroe & Nishimura, 1956

Species	Toxin	Reference
vasinfectum	Lycomarasmin	See *F. o.* f. sp.
	Lycomarasmic acid	*lycopersici*
	Aspergillomarasmin B	
	Fusaric acid	Venkata Ram, 1958
F. poae	Culture filtrate	Bolton & Nuttall, 1968
F. roseum acuminatum	Enniatins	Deol *et al.*, 1978
F. sambucinum	Enniatins A & B	See *F. avenaceum*;
		Audhya & Russell, 1974
F. scirpi	Enniatins A & B	See *F. avenaceum*
F. solani f. sp. *pisi*	Naphthazarins	See *F. javanicum*
F. udum	Fusaric acid	See *F. batatis*
	Dehydrofusaric acid	
F. vasinfectum	Fusaric acid	See *F. batatis*
	Dehydrofusaric acid	
Gibberella baccata (*F. lateritium*)	Enniatins A & B	Gäumann *et al.*, 1960
G. fujikuroi (*F. moniliforme*)	Fusaric acid	Pitel & Vining, 1970
	Dehydrofusaric acid	
	10-hydroxyfusaric acid	
	Moniliformin	Cole *et al.*, 1973
		Lansden *et al.*, 1974
	Gibberellins	Phinney, 1979
	Tyrosol	Turner, 1971
	Isoleucyljasmonic acid	Aldridge *et al.*, 1971

(Shaw, 1981), is produced during or after the stationary phase of growth of a culture. If produced after this time in infected plants lycomarasmin would have no important role in disease development. At best there is no conclusive evidence that lycomarasmin has a role in disease development and it seems most likely that it is not involved in pathogenesis.

Fusaric acid

Fusaric acid (Fig. 5.1) is produced by at least ten species of *Fusarium* mainly belonging to the group Elegans. Some of these species also produce dehydrofusaric acid and 10-hydroxyfusaric acid. Fusaric acid is perhaps the most intensively studied of the toxins produced by *Fusarium* species and its role in pathogenesis has been studied in a number of host pathogen combinations (Gäumann, Naef-Roth & Kobel, 1952; Lakshminarayanan & Subramanian, 1955; Kern & Kluepfel, 1956; Page, 1959; Davis, 1969). However, despite numerous investigations, the significance of the toxin in pathogenesis is still unclear.

Fusaric acid causes excessive water loss in tomato tissue by damaging

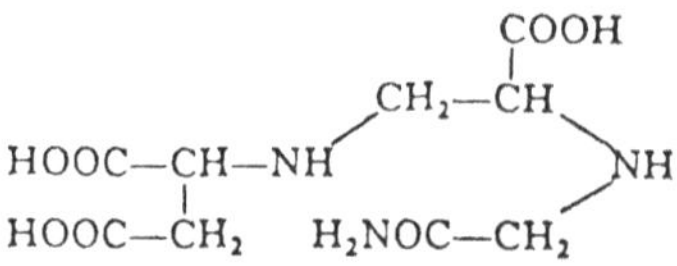

Lycomarasmin

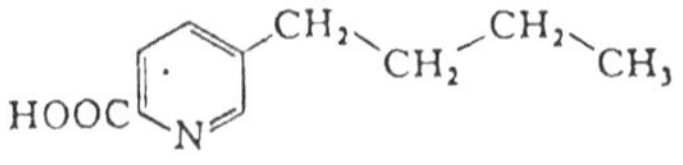

Fusaric acid

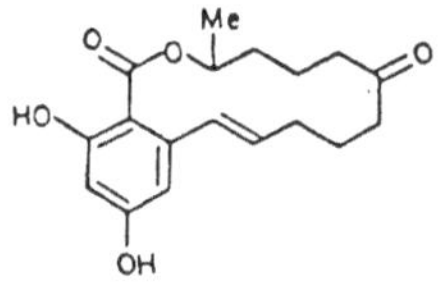

Zearalenone

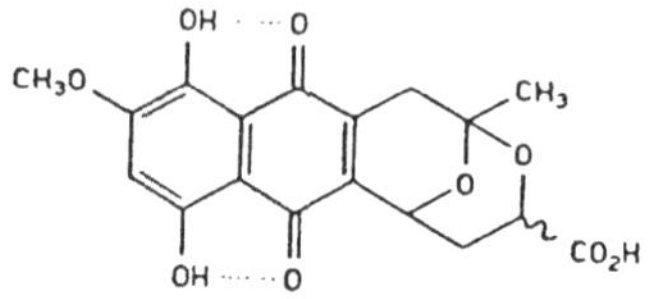

Marticin and isomarticin
α- and β-COOH

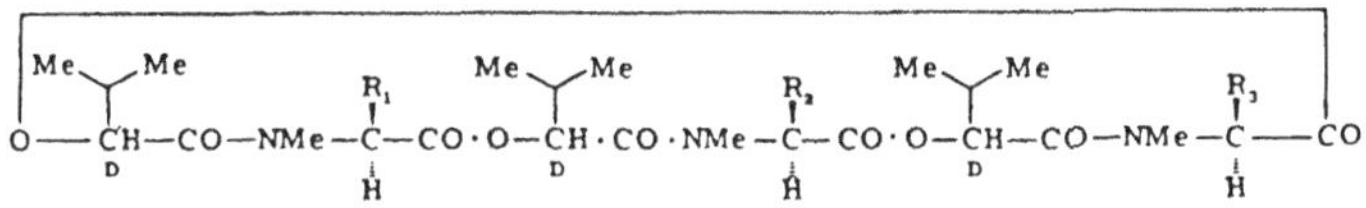

Enniatin A $R_1 = R_2 = R_3 = $ CHEtMe

Enniatin B $R_1 = R_2 = R_3 = $ CHMe$_2$

Fig. 5.1. Toxins produced by species of *Fusarium*.

cell membranes causing leakage of water and other compounds from the cells. The toxin may also have an effect by chelating metal ions and it decreases the respiratory rate in tomato.

Tomato tissue can convert fusaric acid to its *N*-methyl derivative which is much less toxic to plants than the parent compound. Plants resistant to the fungus form more *N*-methyl fusaric acid than is produced by susceptible plants (Kluepfel, 1957; Braun, 1960) but this difference cannot explain the resistance of tomato to *Fusarium* wilt. Pea plants, which are less sensitive than tomato to fusaric acid, also inactivate this acid. Cabbage plants may be the most effective at metabolising fusaric acid but there is some dispute as to whether or not *F. oxysporum* f. sp. *conglutinans*, the forma specialis causing *Fusarium* wilt of cabbage, actually produces the acid in the host (Heitefuss *et al.*, 1960). While fusaric acid has been detected in plants of several host species (e.g. tomato, cotton, banana, flax and watermelon) following inoculation with the appropriate formae speciales of *F. oxysporum*, it has not been detected in cabbage infected with *F. oxysporum* f. sp. *conglutinans*.

With several isolates of *F. oxysporum* f. sp. *niveum* (Davis, 1969) and two isolates of *F. oxysporum* f. sp. *pisi*, Race 2 (Kern, 1972) a correlation has been shown between the virulence of the cultures and the amount of fusaric acid produced both *in vivo* and *in vitro*; however, in other pathogenic strains no such correlation was found. Similarly, in the much more common studies of fusaric acid levels in culture fluids of fungi grown under various conditions, a correlation can be shown in some cases between fusaric acid production and virulence, while in other examples, even involving different isolates of the same species, there is no such relationship.

Studies with UV-induced mutant strains of *F. oxysporum* f. sp. *vasinfectum* (Venkata Ram, 1958) and of *F. oxysporum* f. sp. *lycopersici* (Kuo & Scheffer, 1964) showed that there was no correlation between pathogenicity on the respective hosts and ability to produce fusaric acid in culture. The production of fusaric acid *in vivo* by the mutants was not investigated but the results available suggest that, at least in these two host-pathogen combinations, fusaric acid is not required for pathogenicity.

Fusaric acid inhibits polyphenol oxidase and peroxidase, both of which are involved in the vascular browning reaction characteristic of vascular wilts. Browning is a defence reaction which may be important particularly at early stages of infection. If it does interfere with a host defence mechanism (Wood, 1972) fusaric acid can be regarded as an aggressin, i.e. a specific type of virulence factor rather than a pathogenicity factor (Yoder, 1980).

Zearalenone

Information about this non-host-specific toxin has been summarised by Marré (1980). The compound, an undecenylresorcylic acid lactone (Fig. 5.1; Mirocha, Christensen & Nelson, 1971), first came to notice following development of an oestrogenic syndrome in farm animals which had ingested food containing infected cereals. Only later was the compound recognised as being toxic to plants.

Zearalenone is produced by a number of species of *Fusarium* which are pathogenic to maize or to other cereals. At low concentrations (10^{-5} – 10^{-6} M) zearalenone kills seedlings and inhibits germination (Brodnik, 1975). The information available on the effects of the toxin – depolarisation of membrane potential, electrolyte leakage, inhibition of H^+ and K^+ transport and inhibition of K^+-stimulated membrane ATPase activity (Vianello & Macri, 1978) – suggests that it interacts with the plasma membrane. These effects are manifested in tissues of plants such as beet which are not hosts of the fungus and thus the toxin is not host-specific.

Enniatins

Both the enniatins and the host-specific AM-toxins of *Alternaria mali* are cyclodepsipeptides. The enniatins are hexadepsipeptides with alternating residues of D-2,hydroxyisovaleric acid and branched *N*-methyl amino acids (Fig. 5.1).

At least 12 enniatins have been described from six species of *Fusarium* and a smaller number of fungi belonging to other genera. They are membrane active compounds which chelate metal ions, increase the permeability of membranes to ions (Pressman, 1965) and have a specific effect on the transport of water across membranes (Rudolph, 1976). The toxicity of enniatins to plants presumably derives from these properties and from their action as uncouplers of oxidative phosphorylation (Hunter & Schwarz, 1967). While the structure/activity relationships of the enniatins in relation to their chelating properties and interaction with membranes have been thoroughly investigated (Shemyakin *et al.*, 1969) and they have been isolated from a number of fungi pathogenic to plants in the last five years (Stoessl, 1981), no work on their role in pathogenesis appears to have been published since that of Gäumann's group in 1960 (Gäumann, Naef-Roth & Kern, 1960). While enniatin A and enniatin B are regarded as non-host-specific toxins their similarity to AM-toxins suggests that this problem could be worth re-investigating.

Naphthazarins

Six naphthazarins, 5,8-dihydroxynaphthoquinones with side chains on the C-2 and sometimes on C-3 as well, are known to be produced by species of *Fusarium*. The naphthazarins (Fig. 5.1) are red pigments which readily chelate multivalent metal ions. Marticin and isomarticin are much more toxic to plants than are the other naphthazarins. Kern & Naef-Roth (1967) showed a correlation between the quantity and nature of the naphthazarins produced in culture by *F. solani* f. sp. *pisi* and the virulence of the isolates examined. Marticin and other naphthazarins have been isolated from infected pea tissue in quantities sufficient to damage healthy tissue significantly.

Kern (1972) has suggested that the toxic action of the naphthazarins is due to inhibition of the TCA cycle. Dorn (1974) showed that isomarticin inhibited glutamine synthetase and Roos (1977) has demonstrated that the same compound has a marked effect on the semi-permeability of leaf cells. It is not clear which, if any, of these effects is primarily responsible for the toxic action of the naphthazarins on plants.

References

Aldridge, D. C., Galt, S., Giles, D. & Turner, W. B. (1971). Metabolites of *Lasiodiplodia theobromae*. *Journal of the Chemical Society, C,* 1623–7.

Audhya, T. K. & Russell, D. W. (1974). Natural enniatin A, a mixture of optical isomers containing both *erytho*- and *threo*-N-methyl-L-iso-leucine residues. *Journal of the Chemical Society. Perkin Transactions,* 1, 743–6.

Ballio, A. (1972). Phytotoxins – an exercise in the chemistry of biologically active natural products. In *Phytotoxins in Plant Diseases,* ed. R. K. S. Wood, A. Ballio & A. Graniti, pp. 71–90. London: Academic Press.

Berestetskii, O. A., Nadkernichnyi, S. P. & Patayka, V. F. (1976). Isolation and characteristics of a substance produced by *F. lateritium. Chemical Abstracts,* **84,** 26631.

Bishop, G. C. & Ilsley, A. H. (1978). Production of enniatin as a criterion for confirming the identity of *Fusarium lateritium* isolates. *Australian Journal of Biological Sciences,* **31,** 93–6.

Bolton, A. T. & Nuttall, V. W. (1968). Pathogenicity studies with *Fusarium poae. Canadian Journal of Plant Sciences,* **48,** 161–6.

Braun, R. (1960). Über Wirkungsweise und Umwandlungen der Fusarinsäure. *Phytopathologische Zeitschrift,* **39,** 197–241.

Braun, R. (1963). Pflanzliche toxine. In *Modern Methods of Plant Analysis, Vol. VI,* ed. H. F. Linskens & M. V. Tracey, pp. 219–43. Berlin: Springer.

Brian, P. W., Dawkins, A. W., Grove, J. F., Hemming, H. G., Lowe, D. & Norris, G. L. F. (1961). Phytotoxic compounds produced by *Fusarium equiseti. Journal of Experimental Botany,* **12,** 1–12.

Brodnik, T. (1975). Influence of toxic products of *F. graminearum* and *F. moniliforme* on maize seed germination and embryo growth. *Seed Science & Technology,* **3,** 691–6.

Camporota, P., Trouvelot, A. & Barbier, M. (1973). Contribution à l'étude des marasmines produites par le *Fusarium oxysporum* f. sp. *melonis. Comptes Rendues de l'Academie des Sciences,* **276,** 1903–6.

Clauson-Kaas, N., Plattner, P. A. & Gäumann, E. (1944). Welkurzeugendes Stoffwechselprodukt von *Fusarium lycopersici* Sacc. *Berichte der Schweizerischen botanischen Gesellschaft,* **54,** 523–8.

Cole, R. J., Kirksey, J. W., Cutler, H. G., Doupnik, B. L. & Peckham, J. C. (1973). Toxins from *Fusarium moniliforme*: effects on plants and animals. *Science,* **179,** 1324–6.

Davis, D. (1969). Fusaric acid in selective pathogenicity of *Fusarium oxysporum. Phytopathology,* **59,** 1391–5.

Deol, B. S., Ridley, D. D. & Singh, P. (1978). Isolation of cyclodepsipeptides from plant pathogenic fungi. *Australian Journal of Chemistry,* **31,** 1379–99.

Dimond, A. E. & Waggoner, P. E. (1953). The physiology of lycomarasin production by *Fusarium oxysporum* f. *lycopersici. Phytopathology,* **43,** 195–9.

Dorn, S. (1974), Zur Rolle von Isomarticin, einem Toxin von *Fusarium martii* var. *pisi* in der Pathogenese der Stengel- und Wurzelfaule auf Erbsen. *Phytopathologische Zeitschrift,* **81,** 193–239.

Durbin, R. D. & Steele, J. A. (1979). What art thou, O specificity? In *Recognition and specificity in plant host-parasite interactions*, ed. J. M. Daly & I. Uritani, pp. 115–31. Baltimore: University Park Press.

Egli, T. A. (1969). Untersuchungen über den Einfluss von Schwermetallen auf *Fusarium lycopersici* Sacc. und den Krankheitsverlauf der Tomaten welke. *Phytopathologische Zeitschrift*, **66**, 223–52.

Gäumann, E. (1951). Neuere Erfahnengen mit Welketoxinen. *Experientia*, **7**, 441–7.

Gäumann, E., Naef-Roth, S. & Kern, H. (1960). Zur phytotoxischen Wirksamkeit der Enniatine. *Phytopathologische Zeitschrift*, **40**, 45–51.

Gäumann, E., Naef-Roth, S. & Kobel, H. (1952). Uber Fusarinsaure ein zweites Welketoxin des *Fusarium lycopersici* Sacc. *Phytopathologische Zeitschrift*, **20**, 1–8.

Hardegger, E., Liechti, P., Jackman, L. M., Boller, A. & Plattner, P. A. (1963). Welkstoffe und Antibiotica Die Konstitution des Lycomarasmins. *Helvetica Chimica Acta*, **46**, 60–74.

Heitefuss, R., Stahmann, M. A. & Walker, J. C. (1960). Production of pectolytic enzymes and fusaric acid by *Fusarium oxysporum* f. *conglutinans* in relation to cabbage yellows. *Phytopathology*, **50**, 367–70.

Hiroe, I. & Nishimura, S. (1956). Pathochemical studies on watermelon wilt. Part I. On the wilt toxin phytonivein produced by the causal fungus. *Annals of the Phytopathological Society of Japan*, **20**, 161–216.

Hunter, F. E. & Schwarz, L. S. (1967). Valinomycin. In *Antibiotics I. Mechanism of Action*, ed. D. Gottlieb & P. D. Shaw, pp. 631–5. Berlin: Springer-Verlag.

Kern, H. (1972). Phytotoxins produced by Fusaria. In *Phytotoxins in plant disease*, ed. R. K. S. Wood, A. Ballio & A. Graniti, pp. 35–48. London: Academic Press.

Kern, H. (1978). Les naphthazarines des *Fusarium*. *Annales de Phytopathology*, **10**, 327–45.

Kern, H. & Kluepfel, D. (1956). Die bildung von Fusarinsaure durch *Fusarium lycopersici in vivo* . *Experientia*, **12**, 181–2.

Kern, H. & Naef-Roth, S. (1966). Kupfer, Aluminium und Eisen als krankheitshemmende Faktoren beider Fusskrankheit der Erbsen. *Phytopathologische Zeitschrift*, **57**, 289–97.

Kern, H. & Naef-Roth, S. (1967). Zwei neue, durch Martiella – Fusarien gebildete Naphthazarin – Derivate. *Phytopathologische Zeitschrift*, **60**, 316–24.

Kiss, J., Naef-Roth, S., Hardegger, E., Boller, A., Lohse, F., Gäumann, E. & Pattner, P. (1960). Welkstoffe und Antibiotika 23. Mitteilung uber die Isolierung von Culmonomarasmin, einen peptidartigen Welkstoff aus dem Culturfiltrat von *Fusarium culmorum*. *Helvetica Chimica Acta*, **43**, 2096–101.

Kluepfel, D. (1957). Über die Biosynthese und die Umwandlungen der Fusarinsaure in Tomatopflanzen. *Phytopathologische Zeitschrift*, **29**, 349–79.

Kuo, M. S. & Scheffer, R. P. (1964). Evaluation of fusaric acid as a factor in the development of *Fusarium* wilt. *Phytopathology*, **54**, 1041–4.

Lakshminarayanan, K. & Subramanian, D. (1955). Is fusaric acid a vivotoxin? *Nature*, **176**, 697–8.

Lansden, J. A., Clarkson, R. J., Neely, W. C., Cole, R. J. & Kirksey, J. W. (1974). Spectroanalytical parameters of fungal metabolites. IV. Moniliformin. *Journal of the Association of Official Analytical Chemists*, **57**, 1392–6.

Marré, E. (1980). Mechanism of action of phytotoxins affecting plasmalemma function. *Progress in Phytochemistry*, **6**, 253–84.

Minasyan, A. E., Chermenskii, D. N. & Ellanskaya, I. A. (1978). Synthesis of enniatin B by *Fusarium sambucinum*. *Chemical Abstracts*, **88**, 166485.

Mirocha, C. J., Christensen, C. M. & Nelson, G. H. (1971). F-2 (Zearalenone) estrogenic mycotoxin from *Fusarium*. In *Microbial Toxins VII. Algal and Fungal Toxins*, ed. S. Kadir, A. Ciegler & S. J. Ajl, pp. 107–38. London: Academic Press.

Nishimura, S. (1957). Observations on the fusaric production of the genus *Fusarium*. *Annals of the Phytopathological Society, Japan*, **22**, 274–5.

Page, O. T. (1959). Fusaric acid in banana plants infected with *F. oxysporum* f. *cubense*. *Phytopathology*, **49**, 230.

Phinney, B. O. (1979). Gibberellin synthesis in the fungus *Gibberella fujikuroi* and in higher plants. In *Plant Growth Substances*, American Chemical Society Symposium Series No. 111.

Pitel, D. W. & Vining, L. C. (1970). Accumulation of dehydrofusaric acid and its conversion to fusaric and 10-hydroxyfusaric acids in cultures of *Giberella fujikuroi*. *Canadian Journal of Biochemistry*, **48**, 623–30.

Pressman, B. C. (1965). Induced active transport of ions in mitochondria. *Proceedings of the National Academy of Sciences, USA*, **53**, 1076–83.

Rabie, C. J., Lubben, A., Louw, A. I., Rathbone, E. B. Steyn, P. S. & Vleggaar, R. (1978). Moniliformin, a mycotoxin from *Fusarium fusarioides*. *Journal of Agricultural and Food Chemistry*, **26**, 375–9.

Roos, A. (1977). Zur Physiologie und Pathologie von *Neocosmospora vasinfecta*. *Phytopathologische Zeitschrift*, **88**, 238–71.

Rudolph, K. (1976). Non-specific toxins. In *Encyclopedia of Plant Physiology*, vol. 4, *Physiological Plant Pathology*, ed. R. Heitefuss & P. H. Williams, pp. 270–315. Berlin: Springer-Verlag.

Ruffner, F. (1974). Physiologische Studien auf *Pisum sativum* und *Solanum lycopersicum* im Zusammenhang mit der Entgiftung von Fusarinsaure. *Phytopathologische Zeitschrift*, **79**, 97–129.

Scheffer, R. P. & Briggs, S. P. (1981). Introduction: A perspective of toxin studies in plant pathology. In *Toxins in Plant Disease*, ed. R. D. Durbin, pp. 1–20. London: Academic Press.

Shaw, P. D. (1981). Production and isolation. In *Toxins in Plant Disease*, ed. R. D. Durbin, pp. 21–44. London: Academic Press.

Shemyakin, M. M., Ovchinnikov, Y. A., Antonov, V. K., Vinogradova, E. I., Skrob, A. M., Malenkov, G. G., Evstratov, A. V., Ryalova, I. D., Laine, I. A. & Melnik, E. I. (1969). Cyclodepsipeptides as chemical tools for studying transport through membranes. *Journal of Membrane Biology*, **1**, 402–30.

Singh, G. P. & Husain, A. (1970). Role of toxic metabolites of *Fusarium lateritium* f. sp. *cajani* in the development of Pigeon pea wilt. *Proceedings of the National Academy of Sciences, India*, **40**, 9–15.

Stoessl, A. (1981). Structure and biogenetic relations: fungal nonhost-specific. In *Toxins in Plant Disease*, ed. R. D. Durbin, pp. 110–220. London: Academic Press.

Turner, W. B. (1971) *Fungal Metabolites*. New York: Academic Press.

Venkata Ram, C. S. (1958). Production of ultraviolet induced mutation in *Fusarium vasinfectum* with special reference to fusaric acid synthesis. *Proceedings of the National Institute of Sciences, India*, **23**, 117–22.

Vianello, A. & Macri, F. (1978). Inhibition of plant cell membrane transport phenomena induced by zearalenone (F-2). *Planta*, **143**, 51–7.

Wood, R. K. S. 1972. The development of lesions in plant diseases. *Indian Phytopathology*, **25**, 17–28.

Yoder, O. C. 1980. Toxins in pathogenesis. *Annual Review of Phytopathology*, **18**, 103–29.

6
Fusarium as a biodeteriogen: a case history

J.L.THOMAS

Revlon Health Care (UK) Ltd, Station Road, Shalford, Surrey GU4 8HE, U.K.

Introduction

The biodeterioration of harvested crops and plants by members of the genus *Fusarium* is well established and extremely important economically whereas their significance as purely industrial biodeteriogens has yet to be established. Evidence will be presented in this chapter establishing the species *Fusarium solani* as an important contaminant and biodeteriogen of certain aqueous pharmaceutical suspensions.

Biodeterioration has been defined (Hueck, 1966) as 'any undesirable change in the properties of a material of economic importance brought about by the vital activities of organisms'. Three types of undesirable changes are probable when industrial materials are contaminated with filamentous moulds, such as *Aspergillus*, *Penicillium* and *Fusarium* species (Onions, Allsopp & Eggins, 1981).

(1) Chemical assimilatory deterioration. During this process the mould utilises the material under attack as a food source, and enzymes are liberated which dissolve susceptible components of the material. The nutrients in solution are then absorbed by the mycelium and transported or utilised directly for cellular growth. In pharmaceutical formulations this often means an active ingredient may lose some or all of its potency, or a preservative system its ability to control microbial challenges and/or alteration of physical properties e.g. pH, viscosity and water activity.

(2) Chemical dissimilatory deterioration. This occurs when the contaminating mould releases secondary metabolites, not concerned with the uptake of nutrients. These metabolites may be mycotoxins and such toxic metabolites produced by species of *Fusarium*, *Penicillium* and

Aspergillus cause the majority of naturally occurring mycotoxicoses in man and his domesticated animals. The mycotoxins of *Fusarium* will be dealt with in chapters 9 to 12 but the most important are zearalenone, T-2 toxin and vomitoxin, the last two belonging to a particularly potent group of biologically active compounds known as the trichothecenes (Davis & Diener, 1979). An appreciation of the role of species of *Fusarium* in alimentary toxic aleukia (Joffe, 1965) and of the even wider significance of *Fusarium* toxins has led to the wave of current interest in this genus in the food industry, although toxin production in other industrial products has not been established or reported.

(3) Physical presence within a material or product. Quite simply this means that the undesirable change is the presence of the contaminating mould and applies to both vegetative mycelia and spores.

Fusarium as a biodeteriogen

Evidence of *Fusarium* species behaving as biodeteriogens in pharmaceutical and cosmetic products is rare. That which is available only recognises the presence of *Fusarium* species in materials (Cartwright, 1978; Wedderburn, 1964; Underwood, 1980); whether they act as biodeteriogens as a result of active growth remains undetermined. The main reason for the lack of information concerning *Fusarium* contamination in the pharmaceutical and cosmetic industry must be the difficulty involved in the isolation and identification of *Fusarium* species, evidence of which can be seen in the present day industrial situation where isolated fusaria are only identified to the genus level (Onions, Allsopp & Eggins, 1981).

Historically the importance of the genus *Fusarium* to plant pathologists and mycologists meant that they took on the responsibility for the taxonomy and identification of this large genus. By the early 1930s more than 1000 species of *Fusarium* had been described, though once the concept of variable species was implemented (Wollenweber, 1932; Wollenweber & Reinking, 1935) and understood the number of *Fusarium* species was reduced to 65 with 55 varieties. Later groups of plant pathologists in California (Snyder & Hansen, 1940, 1941, 1945; Snyder & Toussoun, 1965; Toussoun & Nelson, 1968) condensed the number of species still further, eventually reaching the conclusion that only nine species should be employed. Some Fusariologists felt that Snyder and Hansen had oversimplified the group (Joffe, 1974); however, for the uninitiated quality-control microbiologist in industry, the taxonomy involved with the nine species system still provides enough problems to preclude the routine identification of *Fusarium* isolates. The ability of *Fusarium* species for rapid change, variability with substrate and a loss of the ability to sporulate therefore ensures that the taxonomy of the group remains a subject for the expert.

Fusarium in a pharmaceutical product

Fusarium biodeterioration of a pharmaceutical product was first recognised in my laboratory during 1978, when several unopened packs of an aqueous antacid were found to contain a uniform quantity of viable mould contamination. Pour plate counts of diluted samples revealed colony counts on Sabouraud Dextrose Agar (SDA) of *circa* 10^4 colony forming units (c.f.u.) per ml. Microscopic examination of the aqueous phase also revealed the presence of spores, mycelium and germinated spores of, at that time, an unidentified filamentous mould. The product had not indicated the presence of this contaminant when examined two months earlier, at the time of manufacture. As the contaminated samples had been unopened small quantities of the mould must have been present in the product at the time of manufacture, capable of growth and sporulation. Further investigation revealed that many of the infant antacid batches were affected and that the contaminant was potentially a major problem to our manufacturing process.

After some difficulty the mould was identified as *Fusarium solani* (Mart.) Sacc., on the basis of the characteristics of the colony, and macro- and microconidia as presented by Snyder and Hansen and described in the useful Laboratory Guide by Booth (Booth, 1977). The diameter of the colonies on Potato Dextrose Agar (PDA) was greater than 2.5 cm after three days incubation at 31 °C. A single spore inoculum developed a fine network of greyish white hyphae during the first 18 hours which later tended to mass together to form thicker felted areas identified as sporodochia. The latter process was usually accompanied by the production of blue-brown or violet pigments. The macroconidia were sickle-shaped, slightly curved but not beaked, and they were also thick-walled and widest in their upper half (Fig. 6.1). Dimensions ranged from $35–55 \times 4.5–6\,\mu m$ to $45–100 \times 5–8\,\mu m$ according to the number of cross-septa present (normally 2–5). The microconidia were more abundant in the young culture and were elliptical or ovate in shape and apparently produced singly from simple phialides. Chlamydospores were also recognised in submerged vegetative hyphae.

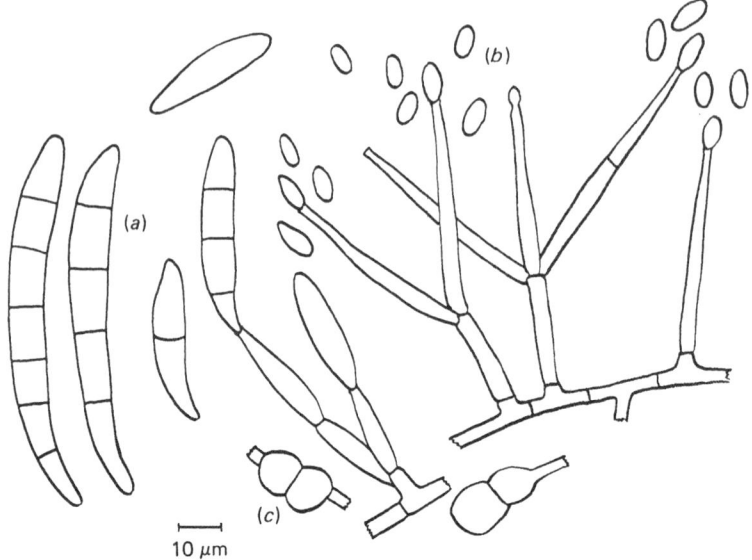

Fig. 6.1. *Fusarium solani* isolated from an infant antacid preparation. (*a*) Macroconidia; (*b*) Microconidia; (*c*) Chlamydospores.

The source of Fusarium solani

An intensive search was mounted to discover the source of this strain of *Fusariurn solani* contaminating the antacid product. Normally such microbial contamination can be traced to the raw materials (Anon., 1971*a*; Westwood & Pin-Lim, 1971), the manufacturing environment or the production personnel (Lennington, 1969). Once the source is identified their introduction into the product can be controlled effectively by using high quality raw materials and by maintaining good manufacturing practices (Anon., 1971*b*). Microbiological examination of the antacid ingredients, including the filtered water supplies (Anon., 1971*c*), continually failed to reveal *Fusarium* species. Studies of the manufacturing environment, however, revealed on one occasion the presence of *Fusarium oxysporum* and *Fusarium solani* in the wash bay area, a significant distance away from the main mixing facilities. Tests in the laboratory, using sterile equipment and laminar flow conditions, to produce the antacid free of *Fusarium* contamination were successful. The evidence therefore suggested that the manufacturing environment was the most likely source, though no direct connexion could be established.

Samples of the contaminated antacid (two months old) were taken and challenged with single inocula of the following microorganisms: *Escherichia coli*, NCTC 10418, *Pseudomonas aeruginosa*, NCTC 6750, *Staphylococcus aureus*, NCTC 10788, *Enterobacter cloacae* and *Citrobacter* species, *Aspergillus niger*, IMI 149007 and *Candida albicans*, NCYC 3179. The antacid formulation was unable to control these potential 'in-use' contaminants, the preservative system of Bronopol (an antimicrobial agent: 2-bromo-2-nitropropane-1,3-diol, 0.0125%) and Potassium sorbate (0.1%) having been effectively neutralised. The product was therefore withdrawn from the market place for reformulation on the following grounds:

(1) That at the time of manufacture the preservative system could not stop the product's biodeterioration by *Fusarium solani*.

(2) That after two months the preservative system was incapable of preventing the colonisation of the product by 'in-use' contaminants (bacteria and fungi). The objectives of the new preservative system were therefore clearly defined in that we had to introduce antifungal properties into the product to defend it from *Fusarium* and other filamentous moulds during manufacture, and that a degree of long term antibacterial and antifungal activity was necessary to prevent in-use contamination so allowing us to fix a reasonable shelf-life to the product.

All these objectives were achieved, the product now being free from all problems associated with *Fusarium* biodeterioration. Studies on the *Fusarium solani* deterioration of the suceptible product continued because of the unusual nature of the problem. The environment within the product implied that growth and sporulation were occurring in submerged culture – a phenomenon not unknown in *F. solani* (Cochrane, 1958). Normally under these conditions sporulation of filamentous moulds is inhibited. The complex nature of sporulation is equally matched by this inhibitory process which involves many factors, such as oxygen tension, changes in the physical nature of the hyphal wall associated with submergence, and the direct contact of sporogenous material with nutrients and inhibitory factors within the respective media or product. Recent studies on the inducement of sporulation in submerged cultures of *Aspergillus nidulans* have shown that before the mould commits itself to sporulate there must be an adequate or fixed quantity of new biomass evolved (Axelrod, 1972); until this level is reached sporulatory stimuli cannot induce sporulation, whereas above that level the mycelium becomes receptive to such stimuli. This point must have been reached in the submerged culture of *Fusarium solani*, resulting in the observed sporulation and cessation of biomass build-up.

The *Fusarium solani* was therefore deteriorating the product by chemical assimilation, fresh biomass being produced at the expense of the antacid suspension. For cell division and enlargement *Fusarium solani* must have an available source of organic carbon (sorbitol, present in the antacid), probably inorganic nitrogen and other trace elements. In *Aspergillus niger* the ratio of carbon to nitrogen has been considered to be one of the factors responsible for the stimulation of the submerged sporulation process (Morton, Dickenson & England, 1960; Morton, 1961); the mould was 'triggered' into the sporulation mode by the exhaustion of its nitrogen source while carbon sources were still available. It now seems likely that trace elements such as calcium and zinc also assist in the induction process (Armstrong *et al.*, 1963). The importance of calcium for sporulation of *P. notatum* and *P. chrysogenum* has also been observed (Hadley & Harrold, 1958). The availability of nitrogen in the antacid suspension was limited due to the inclusion of only one raw material with nitrogen as part of its molecular structure. This source was the antimicrobial agent, Bronopol[R], used in low concentrations (125 mg/l) as the primary preservative system, allowing clean manufacture of the product. Other external sources of nitrogen could have contaminated the product, e.g. water supply, so making the total exhaustion of nitrogen a difficult situation to reach. The ratio of carbon to nitrogen, however, must be similar to that described earlier for work performed with submerged *Aspergillus* cultures; it would seem probable that this ratio encourages a mould, such as *Fusarium solani*, to commit itself to sporulate. Further inducement to sporulate would then arise due to the presence of inorganic metals, such as calcium, zinc, aluminium, silicon etc.

Cappellini & Peterson (1965), working with *Fusarium graminearum*, induced the production of abundant conidia from 'shake cultures' containing a simple salts-yeast extract medium and water-soluble carboxymethylcellulose. Aeration and the substituted cellulose material were found to be critical factors in the sporulation process, aeration being by far the more important. Once again our antacid suspension was seen to have similarities with the media and chemicals used in previous research work, in this case Hypromellose 4000 or hydroxypropylmethylcellulose (HPMC). In the pharmaceutical industry various grades of this water-insoluble colloidal cellulose are used as agents for increasing the viscosity of aqueous suspensions. Increased viscosity slowing down the gravitational 'settling out' process is found with most inorganic suspensions. Our own experiments have demonstrated the effectiveness of HPMC as a sporulatory stimulus for the *Fusarium solani* strain isolated from our antacid product. Earlier workers in this field have all agreed about the essential role aeration has during the submerged sporulation process. Sporulation in our product occurred without this stimulus; however, it is probable that this would have happened earlier if aeration had occurred.

The biodeterioration of the product by masses of *Fusarium solani* spores seems to have been unavoidable, with the carbon–nitrogen ratio, nitrogen exhaustion in the presence of assimilable carbon, trace elements and substituted cellulose material combining to produce an ideal sporulation medium for this particular *Fusarium*.

Degradation of Bronopol by Fusarium solani

As mentioned earlier, the antacid contaminated with *Fusarium solani* was unable to control 'in-use' contamination by potential spoilage bacteria and fungi long before the theoretical breakdown point of the preservative system had been reached. This appears to provide more evidence in support of the hypothesis that *F. solani* is a biodeteriogen assimilating the nitrogen present in the preservative Bronopol, so causing the loss of its ability to maintain the integrity of the product. The nitrogen is probably available to the mould as nitrite derived from the nitro group of Bronopol, the structure of which is shown in Fig. 6.2. Whether nitrite is a preferred nitrogen source for this particular strain of *F. solani* is not known. However, some results noted in unconnected studies have indicated that nitrite may be utilised in preference to nitrate by some microorganisms. Further studies on the inactivation of this preservative have been performed and preliminary data is presented in Fig. 6.3 correlating the rate of deactivation (biodeterioration) of Bronopol with the rate of increase in biomass as measured by dry weight analysis. The activity of Bronopol was measured by microbiological assay using *Pseudomonas aeruginosa* as the test organism. *Fusarium*

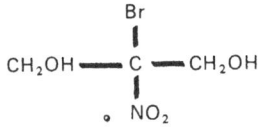

Fig. 6.2. The structure of Bronopol.

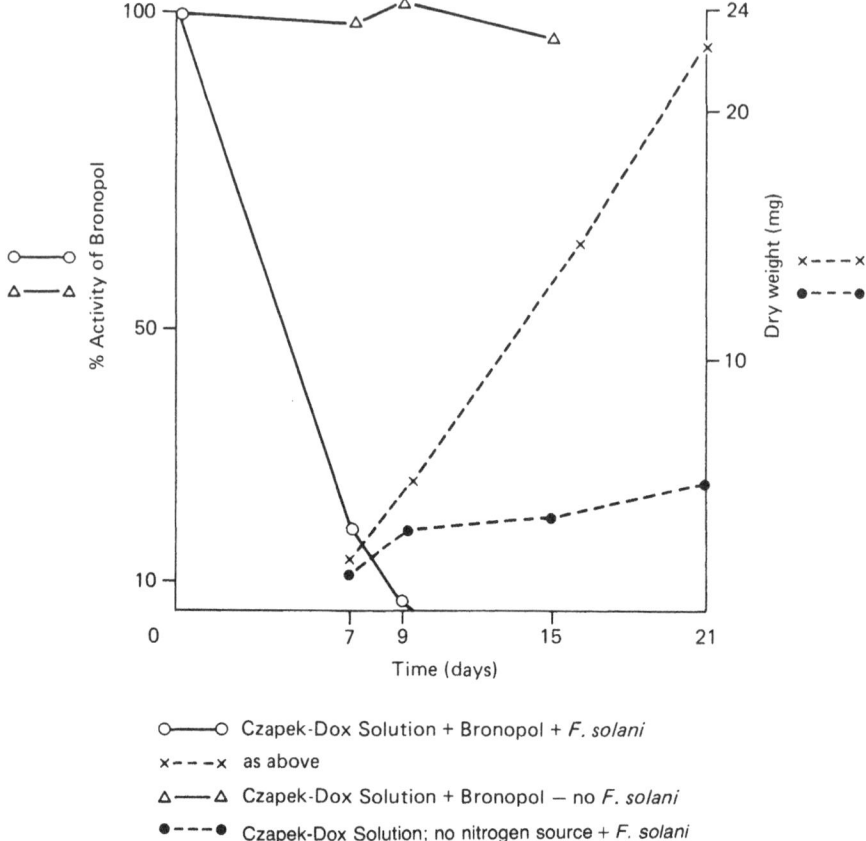

Fig. 6.3. The deactivation of Bronopol by *F. solani*.

solani acts directly on the activity of this nitrogen-containing preservative when recommended 'in use' concentrations are included into minimal media such as Czapek-Dox Solution where the preservative is the sole nitrogen source.

Members of the genus *Fusarium* can thus be directly involved in the biodeterioration (spoilage) of aqueous pharmaceutical suspensions. Their origin can be traced to damp conditions in the manufacturing environments of such products and are therefore not isolated to one specific manufacturing facility. Examination of other antacid suspensions has indicated that this type of contamination, while not common, is certainly not rare. The reformulation of such products to preclude the sporulation process of *Fusarium* would seem to be an important aspect of their future pharmaceutical development.

References

Anon. (1971a). *Guide to Good Pharmaceutical Manufacturing Practice*. London: HMSO.

Anon. (1971b). *Microbiological Quality Control of Pharmaceutical Products and Raw Materials*, vols. I & II. Darmstadt: E. Merck.

Anon. (1971c). Microbial contamination of medicines administered to hospital patients. *The Pharmaceutical Journal*, **207**, 96–9.

Armstrong, J. J., England, D. J. F., Morton, A. G. & Webb, V. A. (1963). Stimulation of sporulation in *Penicillium* by anhydroglucose. *Nature, London*, **197**, 723.

Axelrod, D. E. (1972). Kinetics of differentiation of conidiophores and conidia by colonies of *Aspergillus nidulans*. *Developmental Biology*, **34**, 9–15.

Booth, C. (1977). *Fusarium. Laboratory Guide to the Identification of the Major Species*. Kew, Surrey: Commonwealth Mycological Institute.

Cappellini, R. A. & Peterson, J. L. (1965). Macroconidium formation in submerged cultures by a non-sporulating strain of *Gibberella zeae*. *Mycologia*, **57**, 962–6.

Cartwright, L. G. (1978). Fungal contaminants in pharmaceutical raw materials. *Australian Journal of Pharmaceutical Sciences*, **7**, 61–4.

Cochrane, V. W. (1958). *Physiology of Fungi*. New York: John Wiley.

Davis, N. D. & Diener, U. L. (1979). Mycotoxins. In *Food and Beverage Mycology*, ed. L. R. Beuchat, pp. 397–444. Westport, Connecticut: A. V. I. Publishing Company Inc.

Hadley, G. & Harrold, C. E. (1958). The sporulation of *Penicillium notatum* Westling in submerged liquid culture. 1. The effect of calcium and nutrients on sporulation intensity. *Journal of Experimental Botany*, **9**, 408–17.

Hueck, E. H. (1966). Survey of commercial products used to protect materials against biological deterioration. *International Biodeterioration Bulletin*, **2**, 69–120.

Joffe, A. Z. (1965). Toxin production by cereal fungi causing toxic alimentary aleukia in man. In *Mycotoxins in Foodstuffs*, ed. G. N. Wogan, pp. 77–85. Cambridge, Massachusetts: MIT Press.

Joffe, A. Z. (1974). A modern system of *Fusarium* taxonomy. *Mycopathologia et Mycologia Applicata*, **53**, 201–8.

Lennington, K. R. (1969). Microbiological control. The F.D.A's viewpoint. *Drugs and Cosmetics Industry*, **104**, 44.

Morton, A. G. (1961). The induction of sporulation in mould fungi. *Proceedings of the Royal Society, B*, **153**, 548–69.

Morton, A. G., Dickenson, A. G. F. & England, D. J. F. (1960). Changes in enzyme activity of fungi during nitrogen starvation. *Journal of Experimental Botany*, **11**, 116–28.

Onions, A. H. S., Allsopp, D. & Eggins, H. O. W. (1981). *Smith's Introduction to Industrial Mycology*, 7th ed., pp. 323–46. London: Edward Arnold.

Snyder, W. C. & Hansen, H. N. (1940). The species concept in *Fusarium*. *American Journal of Botany*, **27**, 64–7.

Snyder, W. C. & Hansen, H. N. (1941). The species concept in *Fusarium* with reference to the section Martiella. *American Journal of Botany*, **28**, 738–42.

Snyder, W. C. & Hansen, H. N. (1945). The species concept in *Fusarium* with reference to Discolor and other sections. *American Journal of Botany*, **32**, 657–66.

Snyder, W. C. & Toussoun, T. A. (1965). Current status of taxonomy in *Fusarium* species and their perfect stages. *Phytopathology*, **55**, 833–7.

Toussoun, T. A. & Nelson, P. E. (1968). *A Pictorial Guide to the Identification of Fusarium Species According to the Taxonomic System of Snyder and Hansen.* University Park, Pennsylvania: Pennsylvania State University Press.

Underwood, E. (1980). Ecology of microorganisms as it affects the pharmaceutical industry. In *Pharmaceutical Microbiology*, 2nd edn., ed. W. B. Hugo & A. D. Russell, pp. 253–65. London: Blackwell.

Wedderburn, D. L. (1964). Preservation of emulsions against microbial attack. In *Advances in Pharmaceutical Sciences*, vol. 1, ed. H. S. Bean, A. H. Beckett & J. E. Carless, pp. 195–268. London & New York: Academic Press.

Westwood, N. & Pin-Lim, B. (1971). Microbial contamination of some pharmaceutical raw materials. *The Pharmaceutical Journal*, **207**, 99–102.

Wollenweber, H. W. (1932). Hyphomycetes. *Handbuch der Pflanzenkrankheiten*, Band III, Teil II, pp. 577–819.

Wollenweber, H. W. & Reinking, O. A. (1935). *Die Fusarien.* Berlin: Paul Parey.

7

Fusarium as an insect pathogen

N. CLAYDON* AND J.F. GROVE†

*ARC Unit of Invertebrate Chemistry and Physiology, University of Sussex,
Falmer, Brighton, Sussex BN1 9RQ, U.K.*

Introduction

Entomogenous *Fusarium* species are known to occur in ten out of the 12 sections of the genus, classified according to the scheme proposed by Booth (1971). These species are listed in Table 7.1. Unlike the more widely distributed insect pathogens *Beauveria bassiana* and *Metarhizium anisopliae*, entomopathogenic fusaria show marked host specificity. Of the 15 entomogenous spp., nine are known to be pathogenic, but frequently only one host has been reported for each pathogen.

Because of a long-standing interest in the chemistry of the trichothecenes we examined the possibility of a correlation between entomopathogenicity and the ability to produce trichothecenes. Tables 7.1 and 7.2 show this to be unlikely. There are 20 *Fusarium* spp. known to produce trichothecenes *in vitro* and of these only six are insect pathogens. On the other hand, 11 of the 15 entomogenous fusaria and six out of the nine pathogenic spp. are known producers of trichothecenes. It must, of course, be remembered that not all strains of a pathogenic *Fusarium* species show pathogenicity; and that not all strains of trichothecene-producing *Fusarium* species actually produce trichothecenes *in vivo*, or indeed, *in vitro*.

Tables 7.3 and 7.4 list the trichothecenes produced *in vitro* by certain strains of the six *Fusarium* spp. which are also entomopathogenic. In general, these trichothecenes are some of the more commonly-occurring

* Present address: Glasshouse Crops Research Institute, Worthing Road, Rustington, Littlehampton, West Sussex BN16 3PU, U.K.
† Present address: School of Molecular Sciences, University of Sussex, Falmer, Brighton, Sussex BN1 9QJ, U.K.

Table 7.1. *Distribution of entomogenous and entomopathogenic spp. in the genus* Fusarium

Section		Insect pathogen (P)	Trichothecene producer (+)	Insect	Common name	Stage	Reference
Arachnites	F. nivale		+	Tenebrio molitor†	mealworm	larvae	2
	F. dimerum		+	Scale insects			3
Martiella	F. solani	P	+	Bark beetles		all	4–6
Sporotrichiella	F. poae		+	Oulema gallaeciana	cereal leaf beetle	pupae	7
Arthrosporiella	F. avenaceum		+	Oulema gallaeciana	cereal leaf beetle	pupae	7
	F. semitectum	P		Forest Hymenoptera and Lepidoptera		all	8
Coccophilum	F. larvarum	P		Adelges piceae	Balsam woolly aphid	larvae & adults	9–10
	F. coccophilum	P		Scale insects			1
	F. juruanum			Scale insects			1
Lateritium	F. lateritium	P	+	Oulema gallaeciana	cereal leaf beetle	pupae	7
Liseola	F. moniliforme	P	+	Pyrausta nubialis	maize borer	larvae	11
Elegans	F. oxysporum	P	+	Aedes detritus	mosquito	larvae	12–13
			·	Ostrinia nubilalis	European corn borer	eggs	
Gibbosum	F. equiseti	P	+	Forest Hymenoptera and Lepidoptera		all	8
Discolor	F. sambucinum		+	Oulema gallaeciana	cereal leaf beetle	pupae	7
	F. heterosporum	P	+	Cerambycid beetle		larvae	14

Note: † in vitro test

1. Booth (1971); 2. Davis et al. (1975); 3. Steinhaus & Marsh (1961); 4. Moore (1971, 1973); 5. Barson (1976); 6. Claydon et al. (1977a); 7. Miczulski & Machowicz-Stefaniak (1977); 8. Kalvish (1979); 9. Smirnoff (1970); 10. Fedde (1971); 11. Vago (1958); 12. Hasan & Vago (1972); 13. Lynch & Lewis (1978); 14. Batra & Lichtwardt (1962).

Table 7.2. *Insect pathogenicity and trichothecene production in the genus* Fusarium

	Number
Fusarium spp. (Booth, 1971)	51
Trichothecene producers	20
Entomogenous spp.	15
Insect pathogenic spp.	9
Entomogenous trichothecene producers	11
Insect pathogenic trichothecene producers	6

members of this group of natural products. Exceptionally, *F. hetero-sporum* (Table 7.4) is the source of a number of unusual trichothecenes specific to this organism. The commonly-occurring trichothecenes listed in Table 7.3 were tested (Grove & Hosken, 1975) for toxicity to larvae of the mosquito *Aedes aegypti* in the photomigration bioassay devised by Burchfield, Hilchey & Storrs (1952) (Table 7.5). They were not very active. The most active compound, T-2 toxin, had only 1/25th of the activity of carbaryl whilst, surprisingly, vomitoxin was virtually inactive. This essentially negative result encouraged us to examine some entomo-pathogenic *Fusarium* strains to see if novel insecticidal secondary metabolites of low MW were produced *in vitro*. It was hoped that this research might provide leads for new synthetic insecticides. So far only three of the nine pathogenic spp. have been examined. These are *F. solani* (Claydon, Grove & Pople, 1977a), *F. lateritium* (Grove & Pople, 1980) and *F. larvarum* (Claydon, Grove & Pople, 1979; Grove & Pople, 1979).

Fusarium solani (Mart.) Sacc.

F. solani has been reported to be pathogenic to larvae and adults of the Southern pine beetle *Dendroctonus frontalis* (Moore, 1971, 1973) and to all larval stages of the elm bark beetle *Scolytus scolytus* (Barson, 1976). It was also pathogenic to the blowfly *Calliphora erythrocephala* in an *in vitro* test (Claydon, Grove & Pople, 1977a). On the other hand Kok & Norris (1972) regarded it as an ambrosial fungus and an important source of nutrients for the beetle *Xyloborus ferru-gineus*. Claydon, Grove & Pople (1977a) examined two strains: a strain (47) isolated from *S. scolytus* by Barson was compared with a second (81) obtained by Lightner & Fontaine (1975) from the lobster *Homarus americanus* from an experimental farm in New York. The strains were

Table 7.3. *12,13-Epoxytrichothec-9-enes isolated in vitro from five entomopathogenic* Fusarium spp.

Trivial name	Trichothecene Structure (I)					*F. solani*	*F. oxysporum*	*F. lateritium*	*F. equiseti*	*F. moniliforme*
	R^1 R^2	R^3	R^4	R^5						
Diacetoxyscirpenol	H H	H	OAc	Ac	+	+	+	+	+
Deacetylanguidin	H H	H	OH	Ac	+	+	+	+	+
T-2 toxin	H X	H	OAc	Ac			+	+	+
Diacetylnivalenol	=O	OH	OAc	Ac	+		+	+	
Vomitoxin	=O	OH	H	H				+	+

Note: X = .O.CO.CH₂.CHMe₂.

Table 7.4. *12,13-Epoxytrichothec-9-enes isolated* in vitro *from* F. heterosporum

	Structure (II)
R^1	R^2
Ac	CO.CH$_2$CHMe$_2$
Ac	CO.CH$_2$.C(OH)Me$_2$
H	CO.CH$_2$.C(OH)Me$_2$
Ac	H
H	H

(II)

R^{2}O groups on ring structure; H O H; O; -OH; OH; CH$_2$OR^1

Source: From Cole *et al.*, 1981

Table 7.5. *Larvicidal activity of some 12,13-epoxytrichothec-9-enes to* Aedes aegypti

Common name	Concentration (μg/ml)	Mortality (%) Days: 1	2	3
Diacetoxyscirpenol	25	10	19	49
Deacetylanguidin	25	19	24	26
T-2 toxin	25	27	67	81
Diacetylnivalenol	25	5	8	22
Vomitoxin	100	3	4	4
Carbaryl	1	54	67	73

From Grove & Hosken, 1975.

grown in surface culture on Czapek-Dox Medium and solvent extracts of the culture filtrates were tested for insecticidal activity to *C. erythrocephala*. This activity was wholly accounted for by the formation of the naphthazarin pigments fusarubin (III) and anhydrofusarubin (IV) from both strains, together with javanicin (V) and fusaric acid (VI) from strain 81. The naphthazarin pigments (III)–(V) are well-known phytotoxic metabolic products of *F. solani* and in this respect the two strains were in no way different from phytopathogenic strains isolated from higher plants. When grown in Raulin-Thom medium the yield of the naphthazarin pigments was greater and comparable with that from plant pathogenic strains. The phytotoxin fusaric acid had not previously been reported from *F. solani*. It was completely free from dehydrofusaric acid

Table 7.6. *Insecticidal activity of* F. solani *metabolites to* C. erythrocephala

Compound	Structure	Dose (μg/fly)	Knockdown (% flies down) Hrs: 0	1	4	Mortality (%) Days:1	2	3	48hr LD_{50} (μg/fly)
Fusarubin	III	7.0	40	40	10	50	50	50	7.0
Anhydrofusarubin	IV	0.5	20	20	30	30	30	40	
Javanicin	V	0.5	0	0	0	20	40	55	0.7
Fusaric acid	VI	20.0	93	83	0	47	55	70	18.0

Source: From Claydon *et al.*, 1977.

Table 7.7. *Structure of the components of the enniatin complex*

Enniatin			R	
			CHMeEt	CHMe$_2$
A			3	0
A$_1$	Me R CHMe$_2$		2	1
B$_1$	—N.CH.CO.O.CH.CO—		1	2
B		$_3$	0	3

Table 7.8. *Insecticidal activity of enniatin to* C. erythrocephala

	Dose (μg/fly)	Knockdown (% Flies down)			Mortality (%)	
		Hrs:0	1	3	Days:1	2
Enniatin complex†	10	90	75	55	60	60
Enniatin A	5*	62	26	9	15	32

Source: From Grove & Pople, 1980.
Note: † B : B$_1$: A$_1$ = 5 : 3.5 : 1
* saturated solution.

(VII), a co-metabolite when produced by *F. oxysporum* (Claydon, Grove & Pople, 1977b). The insecticidal activity of the *F. solani* metabolites is shown in Table 7.6. The least polar naphthazarins appeared to be the most toxic.

Fusarium lateritium Nees

F. lateritium is a well-known plant pathogen causing wilt, dieback and canker of woody trees and shrubs. It has been reported to be pathogenic to the cereal leaf beetle (Miczulski & Machowicz-Stefaniak, 1977) and to the scale insect *Hemiberlesia rapax* (L. N. Ferguson, personal communication). Grove & Pople (1980) examined three strains obtained from Ferguson. These strains, when grown on a glucose–peptone medium produced, like the plant pathogenic strains of this *Fusarium* sp., complex mixtures of the cyclic hexadepsipeptide enniatins (VIII) (Table 7.7), in which the proportions of the component enniatins varied from strain to strain. The insecticidal activity of the enniatin complex is shown in Tables 7.8 and 7.9. Against mosquito larvae the enniatins were considerably less active than carbaryl.

Table 7.9. *Larvicidal activity of enniatin to* A. aegypti

	Concentration (μg/ml)	Mortality (%)		
		Hrs: 18	48	72
Enniatin complex†	75*		63	71
Enniatin A	20	25	37	35
Carbaryl	1	54	67	73

Source: From Grove & Pople, 1980.
Note: † B : B_1 : A_1 = 5 : 3.5 : 1
* saturated solution.

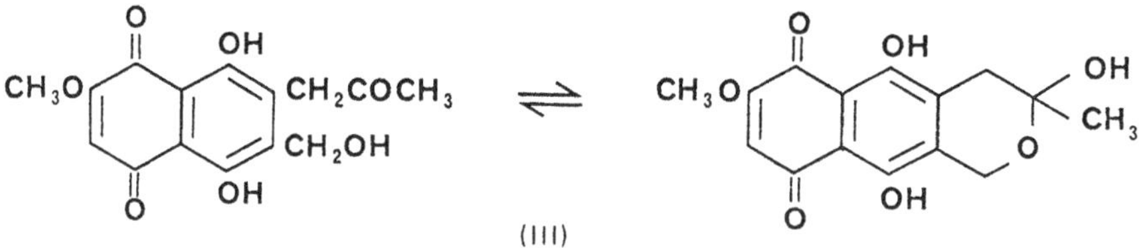

(III)

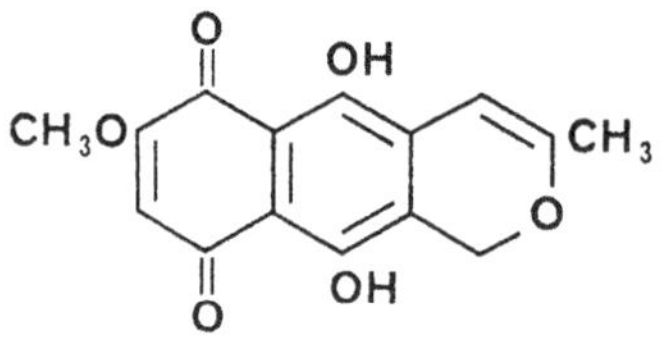

(IV)

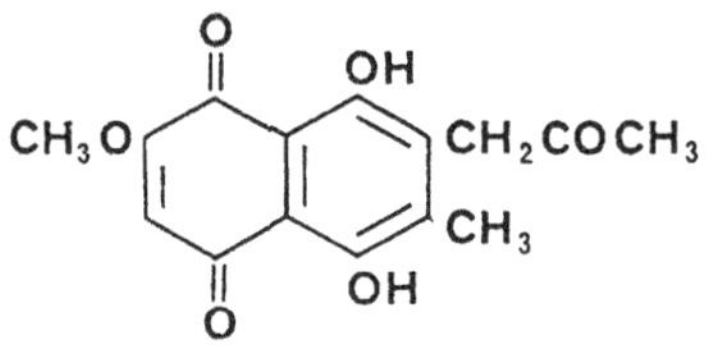

(V)

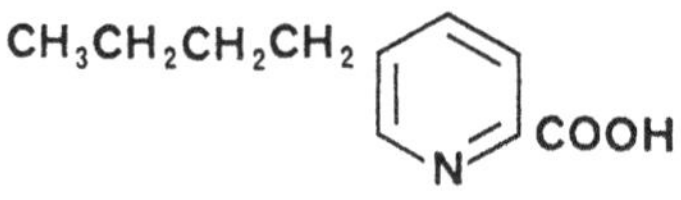

(VI)

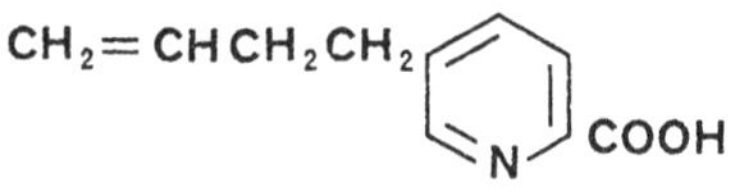

(VII)

Fig. 7.1. Metabolites of *Fusarium solani* and *F. oxysporum*.

Table 7.10. *Insecticidal activity of F. larvarum metabolites to C. erythrocephala*

Compound	Structure	Dose (μg/fly)	Knockdown (% Flies down)				Mortality (%)		
			Hrs: 0	1	3		Days: 1	2	
Fusarentin 6-methyl ether	(IX; R = H)	10*	40	40	10		30	30
Fusarentin 6,7-dimethyl ether	(IX; R = Me)	7	63	20	26		33	36
Monocerin	(X; R = Me)	17.5*	100	77	25		30	37
(+)-Mellein	XI	5.5*	87	70	10			13

Source: From Claydon, Grove & Pople, 1979.
Note: * saturated solution.

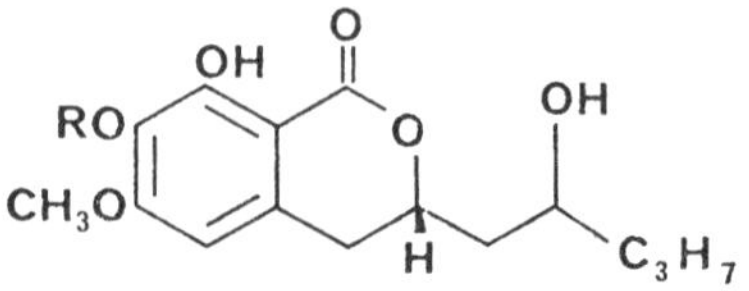

Fig. 7.2. Metabolites of *Fusarium larvarum*.

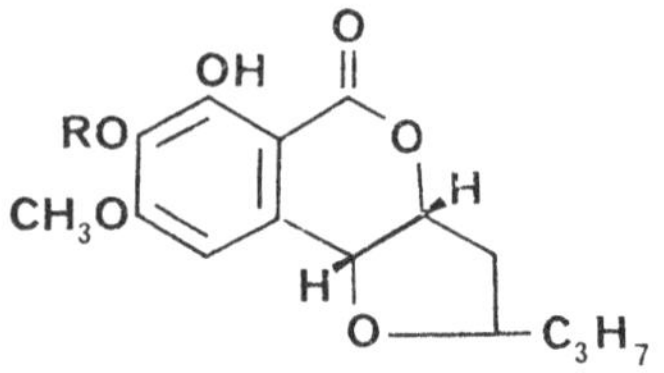

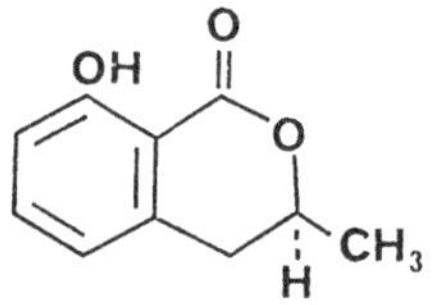

Fusarium larvarum Fuckel

F. larvarum is essentially a parasite of scale insects (Booth, 1971) but there is one report of pathogenicity, to the balsam woolly aphid *Adelges piceae* (Smirnoff, 1970; Fedde, 1971). Claydon, Grove & Pople (1979) looked at three strains of this *Fusarium* sp. Strain 26 was obtained direct from Smirnoff; strain 38 came from the same source but had been deposited with the Commonwealth Mycological Institute; and strain 27 (CMI 141200) had been isolated from a scale insect in Iran. Solvent extracts of the culture filtrates of strains 26 and 27 grown on Raulin-Thom medium, were insecticidal to *C. erythrocephala*. The known fungal metabolic products monocerin (X; R = Me) and (+)-mellein (XI) and the new natural products 7-*o*-demethyl monocerin (X; R = H), fusarentin 6-methyl ether (IX; R = H) and fusarentin 6,7-dimethyl ether (IX; R = Me) were isolated and their structures were determined (Grove & Pople, 1979). Strain 38 did not produce insecticidal extracts or any of these secondary metabolites. *F. larvarum*, in common with many other fungal pathogens, rapidly loses the capacity to produce biologically secondary metabolites – in this case, dihydroiso-coumarins – when maintained in culture on synthetic media. In these circumstances the biochemical similarity of strains 26 and 27, isolated from widely different localitities and hosts, is remarkable. The insecticidal activity of the *F. larvarum* metabolites to *C. erythrocephala* is recorded in Table 7.10. They were, collectively, responsible for all the insecticidal activity present in the extracts.

There is some evidence (Kern & Naef-Roth, 1967) that the naphthazarin pigments are produced *in vivo* in infected higher plants and are responsible, at least in part, for the symptoms associated with phytopathogenicity. Evidence for the production of these compounds, the enniatins and the dihydroisocoumarin metabolites of *F. larvarum*, in infected insects has still to be obtained.

References

Barson, G. (1976). *Fusarium solani*, a weak pathogen of the larval stages of the large elm bark beetle *Scolytus scolytus* (Coleoptera: Scolytidae). *Journal of Invertebrate Pathology*, **27**, 307–9.

Batra L. R. & Lichtwardt, R. W. (1962). Red stain of *Acer negundo*. *Mycologia*, **54**, 91–7.

Booth, C. (1971). *The genus Fusarium*. Commonwealth Mycological Institute, Kew, Surrey.

Burchfield, H. P., Hilchey, J. D. & Storrs, E. E. (1952). An objective method for insecticide bioassay based on the photomigration of mosquito larvae. *Contributions Boyce Thompson Institute*, **17**, 57–85.

Claydon, N., Grove, J. F. & Pople, M. (1977a). Insecticidal secondary metabolic products from the entomogenous fungus *Fusarium solani*. *Journal of Invertebrate Pathology*, **30**, 216–23.

Claydon, N., Grove, J. F. & Pople, M. (1977b). Fusaric acid from *Fusarium solani*. *Phytochemistry*, **16**, 603.

Claydon, N., Grove, J. F. & Pople, M. (1979). Insecticidal secondary metabolic products from the entomogenous fungus *Fusarium larvarum*. *Journal of Invertebrate Pathology*, **33**, 364–7.

Cole, R. J., Dorner, J. W., Cox, R. H., Cunfer, B. M., Cutler, H. G. & Stuart, B. P. (1981). The isolation and identification of several trichothecene mycotoxins from *Fusarium heterosporum*. *Journal of Natural Products*, **44**, 324–30.

Davis, G. R. F., Smith, J. D., Schiefer, B. & Loew, F. M. (1975). Screening for mycotoxins with larvae of *Tenebrio molitor*. *Journal of Invertebrate Pathology*, **26**, 299–303.

Fedde, G. F. (1971). A parasitic fungus disease of *Adelges piceae* (Homoptera: Phylloxeridae) in North Carolina. *Annals of the Entomological Society of America*, **64**, 749–50.

Grove, J. F. & Hosken, M. (1975). The larvicidal activity of some 12,13-epoxy-trichothec-9-enes. *Biochemical Pharmacology*, **24**, 959–62.

Grove, J. F. & Pople, M. (1979). Metabolic products of *Fusarium larvarum* Fuckel. The fusarentins and the absolute configuration of monocerin. *Journal of the Chemical Society (Perkin Trans. 1)*, 2048–51.

Grove, J. F. & Pople. M. (1980). The insecticidal activity of beauvericin and the enniatin complex. *Mycopathologia*, **70**, 103–5.

Hasan, S. & Vago, C. (1972). The pathogenicity of *Fusarium oxysporum* to mosquito larvae. *Journal of Invertebrate Pathology*, **20**, 268–71.

Kalvish, T. K. (1979). Entomophilous fungi of pests from protective forest belts of Kulanda steppe. *Entomology Abstracts*, 6478.

Kern, H. & Naef-Roth, S. (1967). Zwei neue, durch *Martiella*-Fusarien gebildete Naphthazarin-Derivate. *Phytopathologische Zeitschrift*, **60**, 316–24.

Kok, L. T. & Norris, D. M. (1972). Symbiotic interrelationships between microbes and ambrosia beetles. VI. Amino acid composition of ectosymbiotic fungi of *Xyleborus ferrugineus*. *Annals of the Entomological Society of America*, **65**, 598–602.

Lightner, D. V. & Fontaine, G. T. (1975). A mycosis of the American lobster, *Homarus americanus*, caused by *Fusarium* sp. *Journal of Invertebrate Pathology*, **25**, 239–45.

Lynch, R. E. & Lewis, L. C. (1978). Fungi associated with eggs and first-instar larvae of the European corn borer. *Journal of Invertebrate Pathology*, **32**, 6–11.

Miczulski, B. & Machowicz-Stefaniak, Z. (1977). Fungi associated with the cereal leaf beetle. *Oulema gallaeciana* (Coleoptera: Chrysomelidae). *Journal of Invertebrate Pathology*, **29**, 386–7.

Moore, G. E. (1971). Mortality factors caused by pathogenic bacteria and fungi of the southern pine beetle in North Carolina. *Journal of Invertebrate Pathology*, **17**, 28–37.

Moore, G. E. (1973). Pathogenicity of three entomogenous fungi to the southern pine beetle at various temperatures and humidities. *Environmental Entomology*, **2**, 54–7.

Smirnoff, W. A. (1970). Fungus diseases affecting *Adelges piceae* in the fir forest of the Gaspe peninsula, Quebec. *Canadian Entomologist*, **102**, 798–805.

Steinhaus, E. A. & Marsh, G. A. (1961). Report of diagnosis of diseased insects 1951–1961. *Hilgardia*, **33**, 349–479.

Vago, C. (1958). Virulence cryptogamique simultanée vis-à-vis d'un vegetal et d'un insecte. *Comptes rendus Hebdomadaires des Séances de l'Academie des Sciences, Paris*, **247**, 1651–3.

8

Fusarium infections in man and animals

P.K.C.AUSTWICK

Aerobiology Unit, Cardiothoracic Institute, Brompton Hospital, Frimley, Surrey GU16 5QE, U.K.

Introduction

Species of *Fusarium* do not play a very prominent role in human and animal mycosis, and attention to their pathogenic activity is almost entirely through the potential lethal toxigenicity of those species producing trichothecene mycotoxins. It is therefore no surprise that they are only weakly invasive in an infective capacity and generally occur either as colonisers of tissues damaged by other organisms or trauma, or as systemic invaders in hosts who are compromised in their resistance by, for example, immunosuppressive drug treatment. This is not to say that they cannot be virulent primary animal pathogens, as they are in crustacean infections, or that their invasion of non-living or dying tissues does not have serious consequences, as it does in the eye infections which they cause.

Scant attention has been paid to fusarial infections in medical and veterinary mycological texts, but detailed considerations have now appeared by Parker & Klintworth (1971) and Rippon (1982), whilst Rebell (1981) has compiled a very useful chapter with 75 references. To date over 200 instances of human and some 30 of other vertebrate infections are on record, so that 'fusariosis' can now take its place among the 'rarer' mycoses. According to the International Nomenclature of Disease (Anon., 1982), 'fusariomycosis' does not merit sufficient justification to be used as an acceptable disease name, but either of these terms adequately describes the family of syndromes in which *Fusarium* species are primarily or secondarily present.

Fusarium infections may either be overlooked because the characteristic macroconidia are not generally found in or on the infected tissues or in primary culture, or they may be over-reported because of the

Table 8.1. Fusarium *infections in man (number of cases reported)*

Fusarium sp.	Site of infection				
	Eye	Skin	Nail	Subcutaneous	Systemic
*F. episphaeria** (Tode ex Fr.) Snyder & Hansen	1	—	—	—	—
F. dimerum Penzig	5	—	—	—	—
F. moniliforme Sheldon	2	2	—	1	—
F. nivale (Fr.) Ces.	1	—	—	—	—
F. oxysporum Schlecht.	10	5	2	1	3
F. oxysporum var. *redolens* (Wollenw.) Gordon	—	1	1	—	—
F. roseum	—	1	—	—	—
F. solani (Mart.) Sacc.	201	5	1	1	3
F. solani var. *viridiflavum* Ming & Yu	—	1	—	—	—
F. verticillioides (Sacc.) Niren.	1	—	—	—	—
F. spp.	13	—	1	—	3
Totals	234	15	5	3	9

* *F. episphaeria* (Tode) Snyder & Hansen = *F. dimerum* Penzig

growth of contaminating spores derived from the environment which have been superficially deposited on the diagnostic specimens. The isolates of the 22 species of *Fusarium* listed by Wollenweber & Reinking (1935) as 'pathogenic' to vertebrates and invertebrates possibly belong to this category, but no data on them are given. Assessment of the pathogenic significance of any isolate of a *Fusarium* is thus often very difficult, even if good diagnostic material is available. This factor has also meant that the cases from the literature discussed in this account are restricted to those in which there seemed good evidence of tissue invasion in living animals.

Fusarium species infecting man and animals

Tables 8.1 and 8.2 list the *Fusarium* species that have been isolated from clinical specimens and that are apparently causal in infections. *Fusarium solani* is the commonest species represented, chiefly occurring in eye lesions but also isolated from other sites, including the viscera; however, apart from its ubiquitous occurrence in soil and on plants, there seems little reason for its position as the main infecting species. *Fusarium oxysporum* is the second commonest species

Table 8.2. Fusarium *infections in vertebrates (number of cases reported)*

Fusarium sp.	Eye		Skin		Lung		Viscera	
	Mammal	Reptile	Mammal	Reptile	Bird	Reptile	Bird	Reptile
F. equiseti (Corda) Sacc.	—	—	1	—	—	—	—	—
F. moniliforme var. subglutinans (Wollenw. & Reink.)	—	—	1	—	1	—	—	—
F. oxysporum Schlecht.	—	2	2	4	—	—	—	—
F. solani (Mart.) Sacc.	1	1	1	1	—	—	—	—
F. urticearum* (Corda) Sacc.	—	—	—	1	—	—	1	1
F. spp.	3	—	2	4	—	2	—	1
Totals	4	3	7	10	1	2	1	2

*F. urticearum = F. lateritium f. sp. mori (Desm.) Matuo & Sato.

involved in superficial and deep infections, whilst the other identified species have largely been reported from eye infections in man. It is to be expected that many of the unidentified isolates belonged to the species listed, but the names used in compilation could not always be confirmed. The synonyms of a number are given by Booth (1971) and are added to the Tables. The possession of hyalophragmospores enables isolations to be readily placed in the genus, but these spores may themselves be a hazard as Emmons (1944) showed when a culture originally distributed as '*Trichophyton rosaceum*', for use in fungicidal testing of agents against foot ringworm, proved in 10 out of 12 recalled isolates to be *Fusarium oxysporum*.

The species so far reported from human and animal infections have no apparent morphological or physiological features in common which might explain their pathogenicity, despite the finding of Kidd & Wolf (1973) that a keratitis-derived isolate of *F. moniliforme* was capable of a yeast/hypha dimorphism at 25 °C. Their temperature optima and often maxima are below 37 °C but as the mammalian skin surface may be only at 33 °C and the temperature of cold-blooded animals can be near the ambient, there will always be opportunities for growth. However, inability to withstand desiccation is a feature of *Fusarium* hyphae and a very high water content in the substrate is a requirement for almost all species. This may also explain why there is a relatively large number of infections reported in aquatic animals, but does not suggest why some species can cause visceral infection in, admittedly, compromised hosts.

Host species affected

The preponderance of descriptions of fusarial infection in man is understandable, and has provided detailed studies of eye infection in particular; however, the types of infection seen in almost all other sites have also been observed in animals, with the few cases on record probably not reflecting the actual situation. The vertebrate species in which fusarial infections have been reported cover a wide range of animals (Table 8.3), but the main characteristic they have in common is that the individuals affected were all in captivity or domesticated. Whether the stress of confinement could be a predisposing factor in the first group or whether they were observed more closely than in the wild is not clear, but similar infections in free-living animals have not yet been detected. There is also a preponderance of skin infections, especially in aquatic reptiles and mammals, which tends to point to water as a source of these *Fusarium* species. Some 17 of 27 records are

Table 8.3. Fusarium *infections in vertebrate groups (number of cases reported)*

Animal species	Organ affected			
	Skin	Eye	Lung	Viscera
Reptilia				
Tortoises				
Testudo radiata	3	—	—	1
Turtle				
Caretta caretta	1	—	—	—
Lizards				
Chameleon dilepis	1	—	—	—
Lacerta viridis	1	—	—	—
Snakes				
Boa constrictor	—	1	—	—
Epicrates cenchria maurus	—	1	—	—
Indet. snake	1	—	—	1
Alligators & Crocodiles				
Alligator mississipiensis	—	—	1	—
Osteolaemus tetraspis	—	—	1	—
Crocodylus niloticus	1	—	—	—
Amphibia				
African bullfrog				
Pyxicephalus adspersus	1	—	—	—
Mammalia				
Bottle-nosed dolphin				
Tursiops truncatus	2	—	—	—
Baikal seal				
Pusa sibirica	2	—	—	—
Grey seal				
Halichoerus grypus	1	—	—	—
White rhinoceros				
Diceros simus	1	—	—	—
Horse				
Equus caballus	—	3	—	—
Aves				
Avadavit				
Amandava amandava	—	—	1	—
Totals	15	5	3	2

in reptiles, with marine turtles representing the most susceptible group but perhaps also attracting most attention in view of the commercial development of turtle farming. The commonest infections appear to be of newly hatched turtles and their eggshells, and although some may be primary, Rebell (1981) suggests that they may also be associated with herpes infection.

Eye infections

The accessibility of the eye to environmental organisms is probably responsible for the occurrence of the wide range of fungi reported in eye infections. The actual number caused by fungi is probably quite large, with many cases unreported, but more than half of those described have been associated with *Fusarium* species. Almost invariably there has been some damage to the eye, especially the cornea, by abrading or penetrating objects, e.g. soil particles, plant thorns, metal fragments, or by chemicals and heat. These objects may carry *Fusarium* propagules in the form of hyphae, chlamydospores, macro- or microconidia or even possibly ascospores, and although no specific fungal elements have been cited, the work of Basset *et al.* (1965) shows the potential of plant thorns for carrying hyphae (in this case in mycetoma of the foot caused by *Pyrenophora romeroi*).

The conjunctiva is an admirably sticky trapping and retaining surface for airborne fungal propagules which in the normal way are constantly removed by the lachrymal secretions. On conjunctival disruption and exposure of the cornea this cleansing mechanism is interrupted and the propagule may then be caught in the damaged tissue for sufficiently long to germinate. However, it is possible that the main source and transport medium of such propagules is the water used to mechanically wash the offending particles out of the eye. Fusaria are very common aquatic fungi and grow inside domestic and industrial water pipes, tanks and especially taps and, therefore, an inoculum is readily available for the damaged eye tissue which itself may entrap the spores, etc. Emphasis underlining this concept is given by the fact that *F. solani* has been isolated from the majority of the reported cases of *Fusarium* eye infection and at the same time is one of the commonest species obtained from water (Austwick, unpublished data). It was also recovered from the mascara used by a patient with mycotic keratitis by Wilson *et al.* (1971). The need for specially stringent antiseptic precautions to be taken with eye damage is underlined in the cases of fusarial infection which have followed surgery, as reported by Forster (1978). The pathogenesis of these eye infections is difficult to study but there is generally severe inflammation followed by a gradual clouding of the cornea (keratitis) as the fungus colony extends irregularly. This stage can proceed rapidly so that blindness may result in as little as three weeks after the precipitating trauma (Jones *et al.*, 1969). In some cases the cornea itself may be penetrated so that the fungus gains access to the anterior chamber affecting the iris and lens (endopthalmitis).

Keratitis is not uncommon in animals but few cases have been associated with mycotic infection. Three instances are on record of *Fusarium* species infecting the horse (Mitchell & Attleberger, 1973; Peiffer, 1979) and two of *F. oxysporum* infecting snakes. In one of the latter the left eye developed a caseated mass which contained hyphae beneath the horny layer (Zwart *et al.*, 1973) and in the other the lachrymal sac became infected (Vroege, personal communication cited in Zwart *et al.*, 1973).

Skin lesions

Among these is the pustular lesion of the hand of a healthy man described by Collins & Rinaldi (1977), in which *Fusarium moniliforme* hyphae were growing. Hyphae of this species were also seen in the pus of dermal lesions on the flanks of a white rhinoceros, but the fungus may have been a secondary invader (Austwick & Keymer, unpublished). The pustular dermatitis in a dog described by Leinati (1928) in which *F. moronei* (=*F. equiseti* fide Booth) was involved constitutes one of the earliest records of *Fusarium* infection, but Blanchard (1890) recorded the first reptile infection involving a tumour invaded by *F. urticearum* (=*F. lateritium* f. sp. *mori*: see Booth, 1971).

Invasive skin lesions in aquatic mammals provide several records but it is likely that unfavourable water conditions in their display pools contributed to the establishment of infection. *F. oxysporum* was isolated from skin ulcers in two dolphins, and an unidentified species was isolated from lesions on two Baikal seals; in each case extensive penetration of the dermis occurred with minimal inflammatory exposure (Austwick & Keymer, unpublished). Pinnipeds appear especially susceptible to fungal infection of the skin and another case, of papular dermatitis in a grey seal, was associated with *F. solani* by Montali *et al.* (1981). A single case of secondary skin infection by *F. oxysporum* was seen at the London Zoo in 1975 in an African bullfrog which had systemic chromomycosis (due to an unidentified dematiaceous fungus). A white spore-bearing mycelium grew over the head and face of the animal.

According to English (1972), the growth of *F. solani* and *F. oxysporum* in leg ulcers in man is to be regarded as saprophytic and as playing no part in the pathogenesis of the lesions (English, Smith & Harman, 1971). This was also the conclusion reached by Jacobson (1980) in respect of a necrotic dermatitis seen in a python from which *F. solani* had been isolated. Nail infection by *Fusarium* spp. (mainly *F. oxyspor-*

um) is uncommon but similar to that seen with the many saprophytes recorded from cases of onychomycosis of both the toes (Zaias, 1960; McAleer, 1981) and fingernails (Gordon, 1960). Booth (1971) lists an isolate from a foot ulcer as *F. solani* f. sp. *viridiflavum*.

Subcutaneous infection

There are indications that *Fusarium* spp. may be involved in the etiology of 'white grain' mycetoma or madura foot, from the single case reported by Alberici & Marganti (1877) which was first determined as *Acremonium* sp. and is now considered as *F. moniliforme* (see Rebell, 1980). However, according to C. K. Campbell (personal communication) two isolates of *F. solani* and one of *F. oxysporum* from mycetoma lesions are held in the National Collection of Pathogenic Fungi in London.

Systemic infection

All the records of visceral and systemic infection by *Fusarium* spp. in man are in patients whose resistance had been lowered by neoplasia or other debilitating disease, usually combined with immuno-suppressive drug treatment. These include a child with acute leukaemia who developed skin lesions and central nervous system symptoms following *F. solani* infection (Cho *et al.*, 1973) and a man with malignant lymphoma who developed generalised *Fusarium moniliforme* infection (Young *et al.*, 1978). The epidemiological aspects of these infections are put into perspective in the discussion of possible sources of *Fusaria*, e.g. from domestic water taps, and are made more cogent by the contribution in this volume by Thomas (Chapter 6) on *F. solani* contaminating a batch of an antacid preparation. Alimentary tract involvement of fusaria seems to have reached its height in the description by Duncan & Murgatroyd (1938) of a man who presented to them a mycelial cast of his stomach which was later shown to be entirely of *F. oxysporum*.

In animals, isolates from pulmonary deposits seem to indicate infection, but in some cases complications have arisen with the presence of other fungi, especially *Paecilomyces lilacinus* – a known pathogen of reptiles. This fungus holds an unique position in reptilian mycopathology in being isolated from many pulmonary and other lesions but apparently occurring within the caseating tissues as hyphae with very uncharacteristic *Cephalosporium*-like phialides with large (5–11 × 1.8–2.7 μm) ellipsoid conidia. Single phialides certainly occur in some *Paecilomyces* species but these structures appear to be different, so that the concomitant isolation of *Fusarium* and *Cephalosporium* spp. from the same lesions gives rise to questions concerning the identity of the conidiophores in sections (Austwick & Keymer, 1981).

Infection of burns

A small but important group of eight cases of *Fusarium* infection has occurred following severe burns. In one case the infecting *F. roseum* was probably derived from the soil on which the burning child was rolled and Peterson & Baker (1959) considered that the sloughing of the scab crust was actually enhanced by the fungal mycelial covering. In the cases reported by Wheeler *et al.* (1981) and Abramowsky *et al.* (1974), however, dissemination of the fungus to many of the internal organs occurred.

Crustacean infection

Primary infection by fusaria is restricted to invertebrates, especially insects as discussed by Claydon and Grove in this volume (Chapter 7). However recent interest in the farming of both marine and freshwater crustacea has revealed that *F. solani* is an invasive pathogen for the lobster (*Homarus vulgaris*), the crayfish (*Astacus leptodactylus*) and several species of prawn including *Penaeus* and *Pacifastacus* spp. Hard granulomatous lesions have been reported chiefly on the carapace and on the underside of the abdomen and uropod, and invasion of the gills has also been described. The histopathology of the lesions shows mild infiltration by haematocytes but a massive deposition of melanin-containing chitin around the invading hyphae (Alderman, 1981; McAleer, personal communication; Vey, 1978; Bian & Egusa, 1981). Experimental infection was readily obtained by implantation of hyphae in the exoskeleton of crabs but not by injection of spores into the haemocoele (Alderman, 1981).

Experimental infection

An essential demonstration of the invasive pathogenicity of an organism is the fulfilment of Koch's postulates, but this does not always prove possible in the fungi. Thus inoculation of *Fusarium* spp. by the intravenous or intraperitoneal routes in mice is unsuccessful but transient keratitis has been produced in rabbit eyes (Cuero, 1979). Collins & Rinaldi (1977) claimed to have induced lesions by subcutaneous inoculation. In another experiment Adetosoye & Adene (1979) inoculated a *Fusarium* sp. (obtained from a chicken showing nervous symptoms) intracerebrally in 1-day-old chicks and subsequently recovered the fungus from several organs.

Egg infection

Hatching losses in eggs due to infection by *Aspergillus fumigatus* can be extensive when there is external contamination by faeces, but both *F. oxysporum* and *F. culmorum* have also been found in the membranes of dead-in-shell chicks (Ainsworth & Austwick, 1955), and it is probable that there had been invasion of the embryos. Infection by *F. oxysporum* has also been found in a clutch of eggs of an Indian python when extensive blue-green coloration accompanied mycelial growth over the membranes (Austwick & Keymer, 1981).

Discusssion

The ability of certain species of *Fusarium* to invade animal tissue plays an essential if minor part in the biology of these important fungi. They are clearly not major pathogens but their role in eye infections alone demands greater attention because of the rapid and devastating effect of the lesions on the sight of the victim. The few species involved in these infections are known from a very wide variety of habitats both as saprophytes and plant pathogens, but as yet no apparent adaptation to an animal host has emerged. Their only feature which may prove to be of considerable epidemiological importance is their frequent presence in water and on wet surfaces from whence they may be distributed onto damaged tissues or be ingested. Although trauma is an overriding factor, general debilitation of the host by disease, malnutrition, poor housing or immuno-suppressive drug treatment is often contributory, so that when *Fusarium* infections are detected these underlying causes should always be sought. The high proportion of reptiles among the affected animals raises another aspect in view of the reptilian infectivity of several virulent fungal insect pathogens, e.g. *Paecilomyces lilacinus, Beauveria bassiana* and *Metarhizium anisopliae*. Although no experimental inoculations have been reported, it seems likely that certain fusaria, e.g. *F. solani*, will eventually prove to be primary pathogens of reptiles as well. It is to be hoped that the growing medical importance of the fusaria, and their apparent role in juvenile mortality in turtle farms, will stimulate more research on the wider range of infections associated with this genus.

Acknowledgements. I am indebted to Dr Yvonne Clayton, Dr Rose

McAleer, Dr P. M. Stockdale, Dr D. J. Alderman and Professor B. R. Jones for their kind help in the preparation of this chapter.

References

Abramowsky, C. R., Quinn, D., Bradford, W. D. & Conant, N. F. (1974). Systemic infection of *Fusarium* in a burned child. The emergence of a saprophytic strain. *Journal of Paediatrics*, **84**, 561–4.

Abetosoye, A. I. & Adene, D. F. (1978). The response of chickens to experimental infection with *Fusarium* spp. *Bulletin of Animal Health and Production in Africa*, **26**, 198–207.

Ainsworth, G. C. & Austwick, P. K. C. (1955). A survey of animal mycoses in Britain: mycological aspects. *Transactions of the British Mycological Society*, **38**, 369–86.

Alberici, F. & Marganti, L. (1877). Sul primo cas de micetoma geltiele da *Acremonium* sp. osservato in Italia. *Giornale Malettie Instettive Parassitarie*, **12**, 1–4.

Alderman, D. J. (1981). *Fusarium solani* causing an exoskeletal pathology in cultured lobsters *Homarus vulgaris*. *Transactions of the British Mycological Society*, **76**, 25–7.

Anon. (1982). *International Nomenclature of Diseases*. Vol. II, *Infectious diseases. Part 2: Mycoses*, p. 47. Council for International Organizations of Medical Sciences. W.H.O. Geneva.

Austwick, P. K. C. & Keymer, I. F. (1981). Fungi and actinomycetes. In *Diseases of the Reptilia*. vol. 1, eds. J. E. Cooper & O. F. Jackson, pp. 193–231. London: Academic Press.

Basset, A., Caman, R., Baylet, R., & Lambert, D. (1965). Role des épines de Mimosacées dans l'inoculation des mycetomes. *Bulletin Sociétié de la Pathologie Exotique*, **58**, 22.

Bian, B. Z. & Egusa, S. (1981). Histopathology of black gill disease caused by *Fusarium solani* (Martinus) infection in the Kuruma prawn *Penaeus japonicus* Bate. *Journal of Fish Diseases*, **4**, 195–201.

Blanchard, R. (1890). Sur une remarquable dermatose causé chez le lézard vert par un champignon du genre *Selenosporium*. *Memoires de la Société Zoologique de France*, **3**, 241–55.

Booth, C. (1971). *The Genus* Fusarium. Kew: Commonwealth Mycological Institute.

Cho, C. T., Vats, T. S., Lowman, J. T., Brandsberg, J. W. & Josh, F. E. (1973). *Fusarium solani* infection during treatment of an acute leukemia. *Journal of Paediatrics*, **83**, 1928–1031.

Collins, M. S. & Rinaldi, M. G. (1977). Cutaneous infection in man caused by *Fusarium moniliforme*. *Sabouraudia*, **15**, 151–60.

Cuero, R. C. (1979). Distribucion ecologica de *Fusarium solani* y su participacion en la incidencia de la queratitis micotica en Cali, Colombia. *Fitopatologia Colombiana*, **8**, 53–63.

Duncan, J. T. & Murgatroyd, F. (1938). A fungal cast of the stomach vomited. *Transactions of the Royal Society of Tropical Medicine and Hygiene*, **32**, 6–7.

Emmons, C. W. (1944). Misuse of the name 'Trichophyton rosaceum' for a saprophytic *Fusarium*. *Journal of Bacteriology*, **47**, 107–8.

English, M. P. (1972). Observations on strains of *Fusarium solani*, *F. oxysporum* and *Candida parapsilosis* from ulcerated legs. *Sabouraudia*, **10**, 35–42.

English, M. P. Smith, R. J. & Harman, R. R. M. (1971). The fungal flora of ulcerated legs. *British Journal of Dermatology*, **84**, 567–81.

Forster, R. K. (1978). Endopthalmitis. Chapter 24. In *Clinical Opthalmology*, ed. T. D. Duane. Hagerstown, Md.: Harper & Row.

Gordon, W. L. (1960). The taxonomy and the habitats of *Fusarium* species from tropical and temperate regions. *Canadian Journal of Botany*, **38**, 643–58.

Jacobson, E. R. (1980). Necrotizing mycotic dermatitis in snakes: clinical and pathologic features. *Journal of the American Veterinary Medical Association*, **177**, 838–41.

Jones, B. R., Jones, D. B., Lim, A. S. M., Bron, H. A., Morgan, G., Clayton, Y. M. (1969). Corneal and intraocular infections due to *Fusarium solani*. *Transactions of the Opthalmological Societies of the United Kingdom*, **89**, 757–79.

Kidd, G. H. & Wolf, F. T. (1973). Dimorphism in a pathogenic *Fusarium*. *Mycologia*, **65**, 1371–5.

Leinati, F. (1928). Sull'azione patogena di una nuova specie di *Fusarium* (*F. moronei*). *Rivista Biologica*, **10**, 141–54.

McAleer, R. (1981). Fungal infections of the nails in Western Australia. *Mycopathologia*, **73**, 115–20.

Mitchell, J. C. & Attleberger, M. H. (1973). *Fusarium* keratomycosis in the horse. *Veterinary Medicine and Small Animal Clinician*, **68**, 1257–60.

Montali, R. J., Bush, M., Strandberg, J. D., Janssen, D. L., Boness, D. J. & Whitla, J. C. (1981). Cyclic dermatitis associated with *Fusarium* sp. infection in pinnipeds. *Journal of the American Veterinary Medical Association*, **179**, 1198–1202.

Parker, J. C. & Klintworth, G. K. (1971). Miscellaneous uncommon diseases attributed to fungi and actinomycetes. In *Human Infection with Fungi Actinomycetes and Algae*, ed. R. D. Baker, pp. 953–1018. Berlin, Heidelberg, New York: Springer Verlag.

Peiffer, R. L. (1979). Keratomycosis in the horse. *Equine Practice*, **1**, 32–7.

Peterson, J. E. & Baker, T. J. (1959). An isolate of *Fusarium roseum* from human burns. *Mycologia*, **51**, 453–6.

Rebell, G. (1981). *Fusarium* infections in human and veterinary medicine. In Fusarium: *Diseases, Biology, Taxonomy*, ed. P. E. Nelson, T. A. Tousson & R. J. Cook, pp. 212–20. University Park and London: Pennsylvania State University Press.

Rippon, J. W., (1982). *Medical mycology. The Pathogenic Fungi and the Pathogenic Actinomycetes*. 2nd edition, pp. 641–81. Philadelphia: W. B. Saunders.

Vey, A. (1978). Infections fongiques chez l'écrevisse *Astacus leptodactylus* Esch. *Freshwater Crayfish*, **4**, 403–10.

Wheeler, M. S., McGinnis, M. R., Schell, W. A. & Walker, D. H. (1981). *Fusarium* infection in burned patients. *American Journal of Clinical Pathology*, **75**, 304–11.

Wilson, L. A., Kuehne, J. W., Hall, S. W., & Ahearn, D. G. (1971). Microbial contamination in ocular cosmetics. *American Journal of Opthalmology*, **71**, 1298–1302.

Wollenweber, H. W. & Reinking, O. A. (1935). *Die Fusarien*, p. 355. Berlin: Paul Parey.

Young, N. A., Kwong-Chung, K. J., Kubota, T. T., Jennings, A. E. & Fisher, A. (1978). Disseminated infection of *Fusarium moniliforme* drug treatment for malignant lymphoma. *Journal of Clinical Microbiology*, **7**, 589–94.

Zaias, N. (1966). Superficial white onychomycosis. *Sabouraudia*, **5**, 99–103.

Zwart, P., Verver, M. A. J., de Vries, G. A., Hermanides-Nijhof, E. J. & de Vries, H. J. (1973). Fungal infection of the eyes of the snake *Epicrates chenchris maurus*: enucleation under halothane narcosis. *Journal of Small Animal Practice*, **14**, 773–9.

9
Mycotoxicoses associated with *Fusarium*

C.J.MIROCHA

Department of Plant Pathology, University of Minnesota, St. Paul, Minnesota 55108, U.S.A.

Introduction

Species of the genus *Fusarium* are common and widespread in nature, occurring as saprophytes in the soil and in decaying vegetation of all kinds, and as parasites of wild and cultivated plants. It has long been recognised that some of the food grains, particularly barley, rye and wheat, if exposed to wet or humid weather as they mature might be unwholesome. Woronin (1891) described a disease in the eastern U.S.S.R. and Sweden characterised by grain that caused staggers or dizziness ('Taumelgetreide') in humans when such grain was made into bread and consumed. As an example, bread made from mouldy rye resulted in headaches, dizziness, shivering, vomiting, disturbance of vision, and general malaise. Dogs, horses, pigs and chickens were reported as also affected. The disease of cereals which caused the above condition is called *Fusarium* 'scab' or blight and it is still a problem in the world today. As an example, a serious epidemic of 'scab' of wheat occurred in the wheat-growing region of Ontario and Quebec, Canada, in 1980, where up to 20% of the wheat crop was affected. Similarly, the hard red winter wheat of certain regions of Kansas, Missouri and Nebraska in the U.S.A. was hard hit with scab in 1982. The average concentrations of deoxynivalenol in this wheat ranged from 0.2–10.0 p.p.m. but some lots contained 18 or 30 p.p.m. Mainland China has a serious 'scab' problem in wheat each year. Our modern milling practices and technology of bread-making have changed since the time of Naumov (1916) who reported that bread made from wheat and barley infected with *F. graminearum* in Russia (Eastern Siberia) caused inebriation and vomiting. Modern harvesting machines (combines) blow off most of the lighter infected kernels and milling practices separate most of the flour (endosperm) from the infected germ.

In the Hokkaido province of northern Japan, which is climatologically similar to the states of Minnesota and Wisconsin in the U.S.A., Konishi & Ichijo (1970) reported that the feeding of bean hulls to horses caused convulsion, disturbed respiration, decreased heart rate and slowed reflexes. These same authors reported that 10–15% of these horses died. Microbiological examination of the hulls by Ueno *et al.* (1972) led the authors to believe that *Fusarium* toxins may have been involved, but no definitive proof as to etiology was brought forward. Ueno *et al.* (1971) described the red mould disease ('Akakabi-byo') in Japan which is equivalent to the wheat scab found in the U.S.A. and other parts of the world. He described symptoms such as vomiting and diarrhoea in men, refusal of feed in horses, and congestion or haemorrhage in the lung, adrenal, intestine, uterus, vagina and brains of farm animals. Most of these signs have not been reported with 'scabby' wheat ingested by farm animals in the U.S.A. or alternatively veterinarians missed the cause-and-effect relationship during the diagnosis. No signs of vomiting were noted in a human subject who had eaten whole wheat bread made from severely *Fusarium*-infected (scab) wheat (personal experience).

Numerous other diseases suspected of being caused by trichothecenes have been listed such as alimentary toxic aleukia, stachybotryotoxicosis and dendrodochiotoxicosis. In this report we will attempt to identify those mycotoxicoses firmly established as due to intoxication by toxic natural products produced by *Fusarium*. These include intoxications due to T-2 toxin, diacetoxyscirpenol, deoxynivalenol, zearalenone, tibial dyschondroplasia-inducing substance and an antihatchability factor.

Of the more than 40 derivatives of trichothecenes found in laboratory studies (Mirocha, 1979), only five have been found naturally occurring in feedstuff associated with some malady in animals; they are T-2 toxin, deoxynivalenol, diacetoxyscirpenol, monoacetoxyscirpenol and nivalenol (Fig. 9.1*a* and *b*).

Although we tend to describe a certain mycotoxicosis as due to one distinct toxin, it must be remembered that in nature the fungus produces an array of toxins, some known and some as yet undescribed. That is to say: causality is multiple. As an example, Swanson (1980) described the production of the following toxins by one isolate of *Fusarium tricinctum* grown on rice: T-2 toxin (1109 p.p.m.), neosolaniol (400 p.p.m.), HT-2 toxin (91 p.p.m.), acetyl-T-2 toxin (106 p.p.m.), 8-acetylneosolaniol (*a*) Structure of trichothecenes included in Group A:

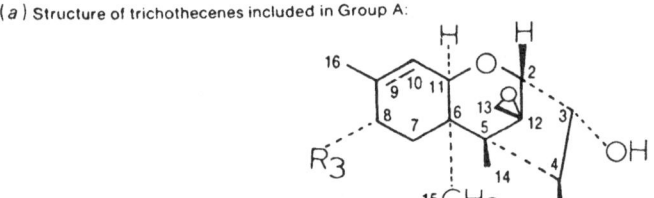

T-2 toxin $R_1 = R_2 = CH_3COO$-, $R_3 = (CH_3)_2CHCH_2COO$-.
HT-2 toxin $R_1 = OH$, $R_2 = CH_3COO$-, $R_3 = (CH_3)_2CHCH_2COO$-;
Neosolaniol $R_1 = CH_3COO$-, $R_3 = OH$, $R_2 = CH_3COO$-;
Diacetoxyscirpenol $R_1 = R_2 = CH_3COO$-, $R_3 = H$;
Monoacetoxyscirpenol $R_1 = OH$, $R_2 = CH_3COO$-, $R_3 = H$.

(*b*) Structure of thrichothecenes included in Group B

T-2 tetraol $R_1 = R_2 = R_4 = OH$, $R_3 = H$
Scirpentriol $R_1 = R_2 = OH$, $R_3 = R_4 = H$.
Deoxynivalenol $R_1 = H$, $R_2 = R_3 = OH$, $R_4 = O$
Fusarenone-X $R_1 = CH_3COO$-, $R_2 = R_3 = OH$, $R_4 = O$.
Nivalenol $R_1 = R_2 = R_3 = OH$, $R_4 = O$

Fig. 9.1. Chemical structure of various trichothecene derivatives.

(22 p.p.m.), TMR-1 (23 p.p.m.), TMR-2 (3 p.p.m.), N-1 (4 p.p.m.), T-2 tetraol (5 p.p.m.) and deacetyl HT-2 toxin (3 p.p.m.).

Alimentary toxic aleukia (ATA)

Within the circles of mycotoxicologists, ATA is almost a household word, often quoted and described as the best-documented account of the effect of *Fusarium* toxins on man resulting from the ingestion of overwintered cereals (millet, wheat, rye, oats, barley and buckwheat) (Joffe, 1978). Graphic accounts of the disease have been described by Joffe (1974) and are summarised here. The ingestion of contaminated grain by man causes a burning sensation in the mouth, oesophagus and stomach. The victim develops a severe case of emesis, diarrhoea and abdominal pain. In the second stage of chronic exposure, there is a marked decrease in leukocytes, and a development of agranulopenia and lymphocytosis. In the third stage, petechial haemorrhages develop on the chest, arms, thigh, and face, and necrotic areas develop in the throat (septic angina). The victim at this point may either die or recover, and if the latter, it may take up to two months for blood-forming capacity of the marrow to return to normal. The disease is also characterised by necrotic angina, sepsis, exhaustion of the bone marrow, bleeding from the nose, throat and gums, and by a fever.

Leonov (1977) reported that ATA has been entirely eradicated in the U.S.S.R. and that discussion of the etiology at this point in time can only be retrospective. He added that: 'It is not possible at the present time to determine what specific compounds of the trichothecene family were the main cause of the human disease that occurred in the U.S.S.R. more than a quarter century ago.' I take exception to this statement, since in fact the causal organism (*Fusarium sporotrichioides*) and T-2 toxin have been shown to reproduce all of the signs of ATA in cats (Yagen *et al.*, 1977; Lutsky *et al.*, 1978; Sato, Ueno & Enomoto, 1975) as well as in man. It had been demonstrated by Joffe (1974) that all of the signs of ATA in man can be duplicated in cats. Moreover, the toxic fraction called poaefusarin (obtained from Soviet scientists) was analysed by Mirocha & Pathre (1973) and found to contain T-2 toxin as the major component. Finally, the signs of ATA have been recognised in man under other circumstances: T-2 toxin has been found to be the major toxic component used in Laos and Kampuchea in chemical warfare directed against the Hmong people (Mirocha *et al.*, 1984). The evidence presented above unequivocally assigns an etiological role of *Fusarium tricinctum* = *F. sporotrichioides* and T-2 toxin in ATA, and truly substantiates ATA as a Fusariotoxicosis. It is very unfortunate that T-2 toxin as well as diacetoxyscirpenol (DAS) should have been tested on man in South-east Asia.

Acute and chronic T-2 toxicosis in swine

In nature, acute toxicity due to ingestion of T-2 toxin is rare. However, laboratory experiments in which single injections were made into the middle vein of the ear of the sow indicated an LD_{50} value of 1.21 ± 0.15 mg/kg body wt. The signs and symptoms include emesis, posterior paresis, lethargy, signs of extreme hunger, rear foot knuckling-over, frequent defecations of normal stools, acute necrosis of many lymphoid tissues and severe mucosal congestion of the jejunum and ileum (Weaver *et al.*, 1978*a*). In evaluation of existing toxicological data, the following signs were confirmed as T-2 attributable lesions: radiomimetic effect (presence of pyknotic nuclei and karyorrhexis) in the small intestine, spleen and lymph nodes, and germinal centres of the lymph follicles of the spleen and lymph nodes in the cat; emetic effect in pigeons and in cats; refusal of the ration in the cow and in swine, rats and pigeons; weakness in cows and hind leg ataxia in cats. The radiomimetic effect was only partial as none was found in the bone marrow but rather was limited to the leukocytes of the spleen, germinal centres of the mesenteric lymph nodes and the lymphoid nodules (Peyer's patches) in the ileum.

Some of the acute signs included periodic muscle tremor, deep and heavy breathing, tachycardia and intensification of the heart beat, extreme pain in the stomach upon palpation and diarrhoea with a large quantity of mucus. Before death, paresis of the hind quarters occurred, characterised by a lack of sensitivity, blue skin and mucous membranes, and a decrease in body temperature.

Swine will readily accept 10–12 p.p.m. of dietary T-2 toxin and can continue on such a diet for up to seven months without any apparent chemical signs of toxicosis. However, at autopsy severe congestion of the bile duct was noted bordering on haemorrhaging.

Acute and chronic toxicity of diacetoxyscirpenol

Diacetoxyscirpenol (DAS) is more potent a toxin than T-2 when administered to swine *via* the middle ear vein. The single dose LD_{50} value is 0.376 mg/kg and the signs associated with toxicity are similar to those of T-2, i.e. emesis, posterior paresis, lethargy, extreme hunger and frequent defecation (Weaver *et al.*, 1978b, 1981).

As an example of the sequence of toxic events, a nonpregnant hybrid sow, 2.5 years old, received 0.43 mg/kg DAS intravenously. On day 1, and within 15 minutes of treatment, the animal vomited frequently, showed extreme hunger by eating its *vomitus* and became lethargic. On day 2, vomition continued; the peripheral middle ear vein (site of injection) collapsed; the animal staggered when attempting to ambulate but did so only when sufficiently prodded; displayed severe posterior paresis and remained prostrate most of the time. On day 3, the animal remained prostrate, when prodded it walked only with the greatest difficulty and accepted some food. On day 4, the animal attempted to stand (due to its own volition) on only its front legs, its hind quarters appeared weak and ataxic. On day 5, the sow moved within its pen very slowly and cautiously; its hind legs occasionally slipped out from under her. The animal appeared to be eating normally. On day 6, the sow showed strong signs of recovery, although unsteady on its legs. The animal was sacrificed on day 7 and the following signs noted at necropsy. The stomach was filled with food; a firm stool was formed in the large bowel; a frothy liquid (medium-yellow in colour) was noted in the lower ileum; severe congestion and possibly haemorrhage was noted in the lower ileum and frank haemorrhaging from the caecum. The myocardium showed acute haemorrhage and congestion.

Dietary DAS fed to growing pigs at 2, 4, 8 and 9 p.p.m. for nine weeks caused multifocal, proliferative, gingival, buccal and lingual lesions. These lesions were not noted in T-2 treated animals. There was a decrease in ration consumption and weight grain and a total refusal at 10 p.p.m. of dietary DAS. The no-effect level appeared to be less than 2 p.p.m. of dietary DAS. The DAS-induced oral trauma seemed to be a contact phenomenon; the optimum combination of toxin concentration and ration contact occurred at 4 p.p.m.

Although DAS caused haemorrhage of the bowel when given intravenously, it did not do so when supplied in the diet at lower concentrations.

Deoxynivalenol (DON) toxicosis

In the harvests of 1980, 1981 and 1982, extensive infection of wheat (*Fusarium* scab) occurred in the U.S., Canada and somewhat less in the United Kingdom. The primary identifiable toxins present in the wheat were deoxynivalenol and zearalenone. The concentrations of deoxynivalenol that caused concern varied between 0.1 p.p.m. and 30 p.p.m. with the average ranging between 2 and 5 p.p.m. It is well-known that the bread baking quality of such wheat is poor but little was known about the toxicity of this wheat to animals and man. The U.S. Food and Drug Administration imposed a 2.0 p.p.m. 'level-of-concern' on all wheat so affected which meant that legally, and if the conditions warranted, such wheat could be seized and impounded. Until the full range of toxins present in 'scabby' wheat is known and their toxicity determined, definitive and meaningful residue tolerances will be difficult to make. The level of concern (2.0 p.p.m.) pertained to the wheat as harvested and not the finished product.

Similarly, maize grown in the U.S. corn belt is also infected by *Fusarium roseum* in the field with concomitant production of deoxynivalenol and zearalenone. These two natural products of *Fusarium* usually occur together as they are products of the same organism. When such corn is fed to swine, it is often partially refused; five per cent infected kernels of a given lot of maize is sufficient to cause refusal.

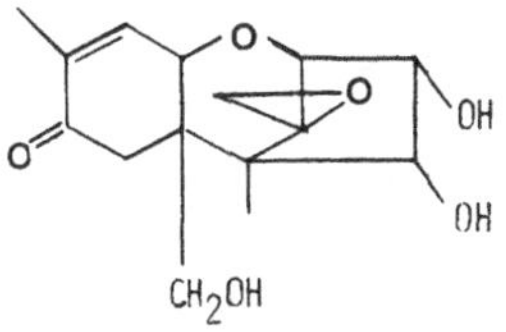

7-dehydroxynivalenol

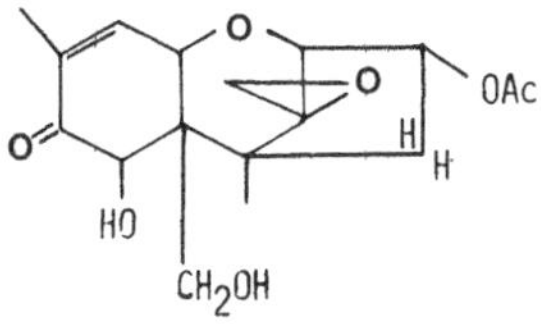

Monoacetoxydeoxynivalenol

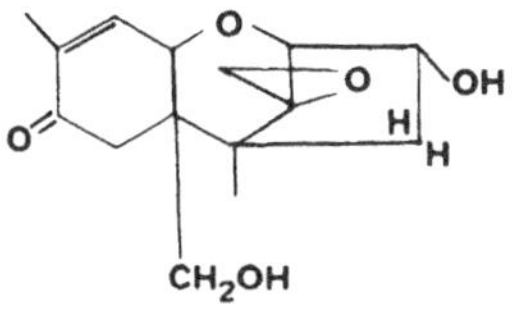

7-dehydroxydeoxynivalenol

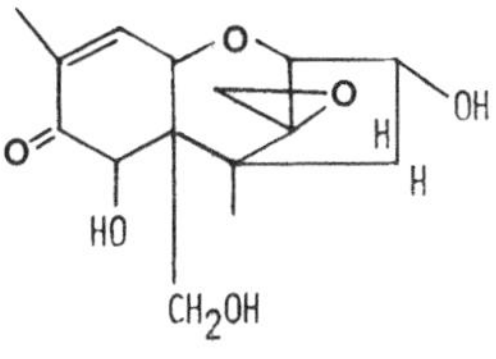

Deoxynivalenol

Fig. 9.2. Monoacetoxydeoxynivalenol (3-acetyldeoxynivalenol) and 7-dehydroxydeoxynivalenol are found naturally occurring in *Fusarium*-infected maize or barley. The derivative (7-dehydroxydeoxynivalenol) is also suspected of being present in *Fusarium*-infected maize.

DON accounts for the major portion of refusal of feed by swine but not for all. Other metabolites such as 3,4,15-trihydroxy-12,13-epoxytrichothec-9-en-8-one (Vesonder *et al.*, 1976) and 7-dehydroxydeoxynivalenol (Bennett *et al.*, 1981) may account for a portion of the refusal activity (Fig. 9.2).

T-2 toxin and DAS are infrequently found in U.S. corn and are almost never found in corn associated with refusal of feed by swine. It is concluded that only those species of *Fusarium* that produce deoxynivalenol are responsible for the naturally occurring maize refused by swine. Naturally occurring trichothecene derivatives resembling deoxynivalenol in polarity, chemistry and biological activity are nivalenol and fusarenon-x (Fig. 9.3).

Fusarium toxins responsible for tibial dyschondroplasia

Not all toxins produced by the many species of *Fusarium* are trichothecenes as exemplified by the butenolide, moniliformin and the phytotoxins fusaric acid and lycomarasmin. Recently, a water-soluble group of toxins (composed of two components) caused a condition in chickens called tibial dyschondroplasia (Walser *et al.*, 1982). Chicks were started on rations containing 2% *F. roseum* grown on rice in culture as well as a partially purified toxin. After a six-week experimental period, the birds at sacrifice showed the presence of a mass

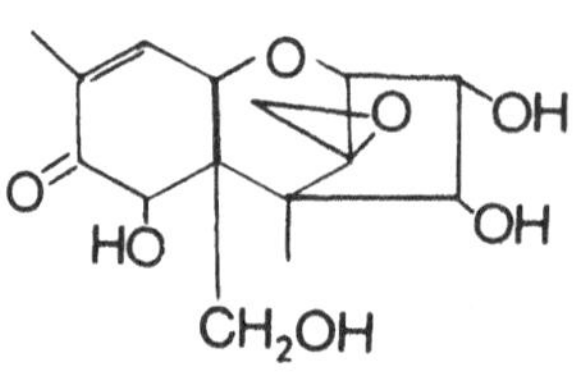

Nivalenol

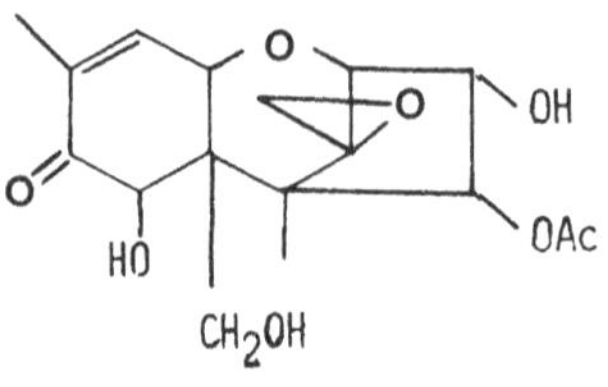

Fusarenon-X

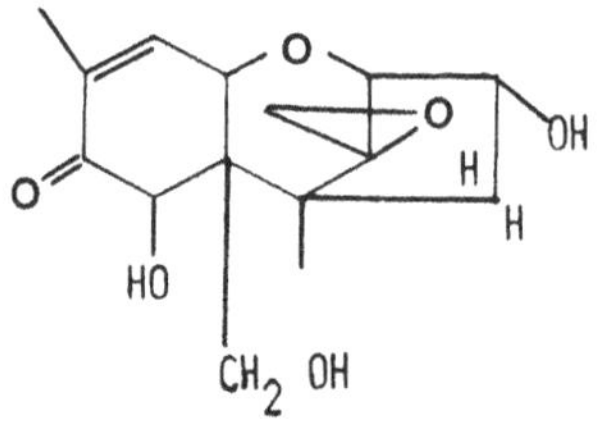

Deoxynivalenol

Fig. 9.3. Similarities in structure of nivalenol, fusarenon-x and deoxynivalenol. The structures suggest a common biogenetic origin.

of cartilage which extended distally from the growth plate on the proximal tibia (Fig. 9.4). Radiographic and histologic findings were typical of osteochondrosis. The number of chondroclasts in 6-week-old chicks with osteochondrosis was less than 10% of those in controls. Defective chondroclasis may contribute to the development of osteochondrosis.

Tibial dyschondroplasia is a common abnormality and not limited to chickens, being also found in turkeys, swine, horses, bovine and dogs. In the past, the causal role had been assigned to rapid weight gain, nutrition, and genetic factors but not to toxins in cereal grains. The discovery that *Fusarium* spp. play a causal role in the etiology of this disease takes it out of the clinician's diagnosis as an idiopathic abnormality. It is important to identify the toxins chemically so that toxins or others similar to them can be looked for as products of fungi and contained in feed.

Toxins responsible for reduction in hatchability in chickens

Poor nutrition or the presence of toxic trichothecenes in diets of poultry can account for reduction in egg production: recently, we have found that some unknown *Fusarium* toxins can also cause a reduction in hatchability, i.e. they do not affect egg production but do kill the

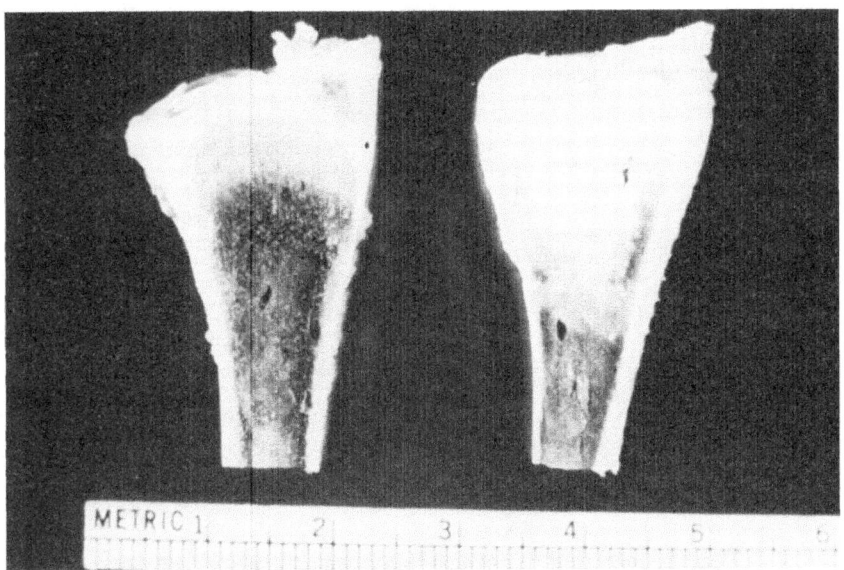

Fig. 9.4. Proximal tibiotarsi from control (left) and from chicks fed dietary *Fusarium* toxins (right). Affected bone has thickened avascular physeal cartilage and abnormal bulbous shape.

embryo. This effect is in contradistinction to the generally toxic effects obtained by feeding dietary levels of diacetoxyscirpenol or T-2 toxin.

A particular isolate of *Fusarium roseum* '*Graminearum*' (Alaska 2-2) was grown on a rice medium in the laboratory and then incorporated into the diet of laying hens (White Leghorn females) at a concentration of 3% of the diet. Birds were inseminated weekly with semen from males given normal diets. An additional group of birds was placed on a diet containing 0.5 p.p.m. diacetoxyscirpenol (DAS) to serve as a positive control. In four weeks, hatchability of fertile eggs was reduced 24% by DAS and 99% by the *Fusarium* culture (Table 9.1). Hatchability returned to normal after the toxins were removed. DAS was only partially responsible for the reduction whereas the major portion of toxicity was due to some other toxins (Allen *et al.*, 1982).

More recent unpublished information leads us to conclude that the same toxins responsible for tibial dyschondroplasia also cause reduction of hatchability. The role of these toxins is subtle in the sense that they are not acutely toxic but affect the embryo and its development gradually. The majority of the embryo mortalities occur prior to the 7th day of egg incubation.

Table 9.1. *Effect of feeding crude* Fusarium roseum 'Grami-nearum' *cultures and diacetoxyscirpenol (DAS) on hatchability of White Leghorn females*

	Week			
	1	2	3	4
Treatments	Per cent hatchability			
Control	92.0[a]	92.7[a]	92.0[a]	95.0[a]
3% culture	63.4[b]	8.0[b]	3.3[b]	3.0[c]
0.5 p.p.m. DAS	88.7[a]	90.0[a]	80.0[a]	58.0[b]

[a, b, c] Means within a column with different superscripts differ significantly from one another ($p < 0.05$).

Zearalenone and hyperoestrogenism

Zearalenone is the trivial name of 6-(10-hydroxy-6-oxo-*trans*-1-undecenyl)-β-resorcylic acid-μ-lactone (see Fig. 11.1). The name is partially derived from the Latin binomial for maize and its lactone and ketone functions. The naturally occurring derivatives of this group of phytoestrogens are: alpha and beta isomers of zearalenol, 8'-hydroxyzearalenone, 6'-8'-dihydroxyzearalene, 3'-hydroxyzearalenone and 7'dehydrozearaldienone. The alpha derivative of zearalanol is used as a growth promoter in animal production and also as a chemotherapeutant in alleviation of post-menopausal discomfort in women.

Zearalenone and the zearalenol isomers can cause serious reproductive problems when consumed by swine or dairy cattle but have little to no effect when consumed by laying hens, broilers and turkeys. Concentrations in the diet as high as 800 p.p.m. have little effect on poultry.

Swine are most sensitive to this oestrogen and when zearalenone, found in infected cereals, is ingested it can cause hyperoestrogenism and infertility. The oestrogenic syndrome can be characterised by swelling of the genital system; in the prepubertal gilt (Fig. 9.5) the vulva becomes swollen and oedamatous and in severe cases may lead to vaginal or anal prolapse; the uterus becomes enlarged, oedematous and tortuous, with atrophy of the ovaries. Young males may experience atrophy of the testes and enlargement of the mammary glands. Zearalenone by itself does not cause abortion in swine but is often found in feeds together with the trichothecenes that have the potential for inducing abortion.

Fig. 9.5. Swollen oedematous vulvae of prepubertal gilts resulting from consumption of zearalenone-contaminated maize.

Moreover, the trichothecenes can also contribute to infertility, thus further compromising reproduction in animals.

The effect of zearalenone on the uterus weight in young gilts given different dosages is summarised in Table 9.2. Uterus weight was increased by administration *per os* of 1 mg or more per day for 8 days. Consumption by the gilts of 80 g/day of *Fusarium*-invaded corn resulted in uterus weight gain equal to that produced by an amount of zearalenone of between 5 and 10 mg per day. This toxin can also cause reduced litter size, weak piglets and pseudo-pregnancy in swine (Chang, Kurtz & Mirocha, 1979).

Dietary zearalenone at a concentration of 100 p.p.m. and consumed by swine for one week caused degenerative lesions in the ovary and the uteral horn (Vanyi, Szeky & Rumvaryne, 1974). Ovulation was suppressed and the degeneration of the uterine glands in the mucosa of the uterus contributed to infertility. When zearalenone was withdrawn, the vulvar swelling gradually disappeared, but the lesions in the reproductive organs remained, i.e. the clinical signs of oestrogenism regressed but the histopathological lesions remained.

Table 9.2. *Effect of F-2 and* Fusarium-*invaded corn on the uterus of gilts*[a]

Treatment	Weight of gilt[b] (kg)	Weight of uterine horn (g)	
		Total	Per kg body weight
Control, sacrificed after 5 days	34.1	27.8	0.81
Control, sacrificed after 14 days	75.5	28.0	0.37
Oestradiol, 0.5 mg IM daily for 4 days	60.5	173.1	2.86
Oestradiol, 0.5 mg IM daily for 14 days	81.4	241.6	2.97
F-2, 1 mg daily for 8 days	49.5	47.0	0.95
F-2, 5 mg daily for 8 days	57.7	85.4	1.48
F-2, 10 mg daily for 5 days	53.6	174.6	3.26
F-2, 25 mg daily for 5 days	76.3	271.6	3.56
F-2, 50 mg daily for 4 days	78.6	251.9	3.20
Fusarium-invaded corn, 80 gm daily for 6 days	59.5	136.9	2.30
Fusarium-invaded corn, 80 gm daily for 14 days	75.5	206.0	2.73

[a] After administration *per os.*
[b] The proprietary material oestradiol (ECP, Upjohn Co.) was administered by intramuscular injection (IM); F-2 by gelatin capsule; *Fusarium*-invaded feed was mixed with normal ration. Prepubertal pigs about 6 weeks of age.

Conclusions and discussion

Some of the mycotoxicoses which can be identified as fusario-toxicoses are: alimentary toxic aleukia, hyperoestrogenism (zearalenone) and refusal of feed (deoxynivalenol). Although T-2 toxin and diacetoxyscirpenol can account for some of the signs and symtoms of various fusariotoxicoses, we do not find them in nature frequently enough to assign causality safely or definitively. Part of the reason for this is that errors due to sampling are large and feed analyses often unreliable. It is suggested that veterinary diagnosticians should use blood, urine and faeces in the analyses of the mycotoxins in order to determine intoxication. This is possible and reasonable with T-2 and zearalenone toxins; the metabolism of diacetoxyscirpenol in animals must be studied in order to determine the feasibility of such analyses for this toxin.

Zearalenone and deoxynivalenol are frequently encountered in cereal grains and perhaps are the best examples of *Fusarium* toxins that affect farm animals. The major signs or effects of Fusarium intoxication in bovine and porcine species are hyperoestrogenism and infertility. In poultry, the proliferative caseous-like plaques in their mouth-parts are diagnostic signs.

T-2, DAS, and DON, if administered properly to animals, will cause haemorrhaging. Other water-soluble components of *Fusarium* cultures cause tibial dyschondroplasia in poultry and prevent hatchability of fertile eggs of chickens.

The incidence of *Fusarium* toxins actually found naturally occurring in feed or foodstuff is small. Toxins or natural products like zearalenone (the phytoestrogen) and the trichothecene called deoxynivalenol are easily found in maize and wheat as well as other cereal grains; they cause **hyperoestrogenism** and **refusal of feed,** respectively. Although trichothecenes such as T-2 toxin and diacetoxyscirpenol are well known, their incidence in nature is infrequent and at least in the U.S.A. not found frequently enough to account for the many suspect fusariotoxicoses reported. This is not to say that *Fusarium* is not associated with the disease, but simply that we cannot account for all of the signs seen in the animals.

I would like to stress that causality is multiple and that in nature animals are not insulted by single pure toxins. We do not always find T-2 or DAS in our problem animal feeds but we do find some of the signs and symptoms that these toxins produce. Are there other attendant mycotoxins in the biological mixture that we should be aware of, or does our method of sampling and analysis need improvement?

References

Allen, N. K., Jevne, R. L., Mirocha, C. J. & Yin Wan Lee. (1982). The effect of *Fusarium roseum* culture and diacetoxyscirpenol on reproduction of White Leghorn females. *Poultry Science,* **61,** 2172–5.

Bennett, G. A., Peterson, R. E., Plattner, R. D. & Shotwell, O. L. (1981). Isolation and purification of deoxynivalenol and a new trichothecene by high pressure liquid chromatography. *Journal of the American Oil Chemists Society,* **38,** 1002A–5A.

Chang, K., Kurtz, H. & Mirocha, C. J. (1979). Effects of the mycotoxin zearalenone on swine reproduction. *American Journal of Veterinary Research,* **40,** 1260–7.

Joffe, A. Z. (1974). Toxicity of *Fusarium poae* and *F. sporotrichioides* and its relation to alimentary toxic aleukia. In *Mycotoxins,* ed. I. F. H. Purchase, pp. 229–262. New York: Elsevier Scientific Publishing Co.

Joffe, A. Z. (1978). *Fusarium poae* and *F. sporotrichiodes* as principal causal agents of alimentary toxic aleukia. In *Mycotoxic Fungi, Mycotoxins, Mycotoxicoses: An Encyclopedic Handbook*, vol. 3, ed. T. D. Wyllie & L. G. Morehouse, pp. 21–86. New York: Marcel Dekker, Inc.

Konishi, T. & Ichijo, S. (1970). *Research Bulletin of Obihiro University*, **6**, 242–58. (In Japanese.)

Leonov, A. N. (1977). Current view of the chemical nature of factors responsible for alimentary toxic aleukia. In *Mycotoxins In Human and Animal Health*, ed. J. V. Rodricks, C. W. Hesseltine & M. A. Mehlman, pp. 323–8. Illinois, U.S.A.: Pathotox Publishers Inc.

Lutsky, I., Mor, N., Yagen, B. & Joffe, A. Z. (1978). The role of T-2 toxin in experimental alimentary toxic aleukia: a toxicity study in cats. *Toxicology and Applied Pharmacology*, **43**, 111–24.

Mirocha, J. (1979). Trichothecene toxins produced by *Fusarium*. In *Conference on Mycotoxins In Animal Feeds And Grains Related To Animal Health*. U.S. Dept. Commerce PB-300-300.

Mirocha, C. J. & Pathre (1973). Identification of the toxic principle in a sample of poaefusarin. *Applied Microbiology*, **26**, 719–24.

Mirocha, C. J., Pawlosky, R. A., Chatterjee, K., Watson, S. & Hayes, W. (1984). Analysis for *Fusarium* toxins in various samples implicated in biological warfare in South-east Asia. *Journal of the Association of Official Analytical Chemists*. (In press.)

Naumov, N. A. (1916). Intoxicating bread. Pliany Khlieb. *Trudy Biuro po Mika i Fitopat.* No. 12, pp. 1–216, Petrograd.

Sato, N., Ueno, Y. & Enomoto, M. (1975). Toxicological approaches to the toxic metabolites of Fusaria. VIII. Acute and subacute toxicities of T-2 toxin in cats. *Japanese Journal of Pharmacology*, **25**, 263–70.

Swanson, S. P. (1980). Characterization of T-2 toxin metabolites in cultures of *Fusarium tricinctum*. Unpublished M.S. Thesis, Dept. Plant Pathology, University of Minnesota, St. Paul, MN 55108, U.S.A.

Ueno, Y., Ishii, K., Sakai, K., Kanaeda, S., Tsunoda, H., Tanaka, T. & Enomoto, M. (1972). Toxicological approaches to the metabolites of *Fusaria. IV.* Microbial survey on 'Bean-hulls poisoning of horses' with the isolation of toxic trichothecenes, neosolaniol and T-2 toxin of *Fusarium solani* M-1-1. *Japanese Journal of Experimental Medicine*, **42**, 187–203.

Ueno, Y., Ishikawa, Y., Nakajima, M., Sakai, K., Ishii, K., Tsunoda, H., Saito, M., Enomoto, M., Ohtsubo, K. & Umeda, M. (1971). Toxicological approaches to the metabolites of Fusaria. I. Screening of toxic strains. *Japanese Journal of Experimental Medicine*, **41**, 257–72.

Vanyi, A., Szeky, A. & Rumvaryne, S. E. (1974). Fusariotoxicoses. V. The effect of F-2 toxin on the sexual activity of female swine. *Magyar Allatorvosok Lapja*, **29**, 723–9. (In Hungarian).

Vesonder, R. F., Ciegler, A., Jensen, A. H., Rohwedder, W. K. & Weisleder, D. (1976). Co-identity of the refusal and emetic principle from *Fusarium*-infected corn. *Applied and Envioronmental Microbiology*, **31**, 280–5.

Walser, M. N., Allen, N. K., Mirocha, C. J., Hanlon, G. F. & Newman, J. A. (1982). *Fusarium*-induced osteochondrosis (Tibial Dyschondroplasia) in chickens. *Veterinary Pathology*, **19**, 544–50.

Weaver, G. A., Kurtz, H. J., Bates, F. Y., Chi, M. S., Mirocha, C. J., Behrens, J. C. & Robison, T. S. (1978a). Acute and chronic toxicity of T-2 mycotoxin in swine. *Veterinary Record*, **103**, 531–5.

Weaver, G. A., Kurtz, H. J., Mirocha, C. J., Bates, F. Y. & Behrens, J. C. (1978b).

Acute toxicity of the mycotoxin diacetoxyscirpenol in swine. *Canadian Veterinary Journal,* **19,** 267–71.

Weaver, G. A., Kurtz, H. J., Bates, F. Y., Mirocha, C. J., Behrens, J. C. & Hagler, W. M. (1981). Diacetoxyscirpenol toxicity in pigs. *Research in Veterinary Science,* **31,** 131–5.

Woronin, M. (1891). Ueber das 'Taumelgetreide' in Sud Ussurien. *Botanische Zeitung,* **6,** 83–94.

Yagen, B., Joffe, A. Z., Horn, P., Mor, N. & Lutsky, I. I. (1977). Toxins from a strain involved in ATA. In *Mycotoxins In Human and Animal Health,* ed. J. V. Rodericks, C. W. Hesseltine & M. A. Mehlman, pp. 329–36. Illinois, U.S.A.: Pathotox Publishers Inc.

10
The natural occurrence of
Fusarium mycotoxins

J.E.SMITH, I.MITCHELL AND M.L.C.CHIU

Department of Bioscience and Biotechnology, Applied Microbiology Division, Strathclyde University, 204 George Street, Glasgow G1 1XW, U.K.

Introduction

The infectious nature of fungi to man, animals and plants has long been recognised. These infections or mycoses can cause specific changes in the host organism exemplified by inflammation, sickness and often death (Emmons *et al.*, 1977). Chronic exposure to certain types of fungal spores may result in allergic responses in sensitised animals, including man, resulting in illness and decreased productivity in animals. The health of animals and man can be further affected by toxic compounds produced by specific toxigenic mould fungi. These compounds have been termed 'mycotoxins' and the resultant diseases as 'mycotoxicoses'.

Mycotoxicosis is poisoning by the ingestion of toxins of fungal origin in foods which have been altered or damaged by the growth of a limited range of toxigenic fungi. Chemically, mycotoxins are relatively low molecular weight, non-antigenic secondary metabolites capable of causing a toxic effect in man and animals. Such compounds are synthesised generally, but not always, on grains and other plant materials, entering the food chain by direct or indirect means (Jarvis, 1976; Moss, 1977) (Table 10.1). Mycotoxins can enter the system of an animal by ingestion, inhalation or direct skin contact and minute quantities can cause significant health changes in exposed animals. At high concentrations many mycotoxins can produce immediate acute disease syndromes while at lower dose levels they can be carcinogenic, mutagenic, teratogenic or oestrogenic. They can also reduce the growth rate of young animals and may even interfere with native mechanisms of resistance and impair immunologic responsiveness, making animals, in particular

Table 10.1. *Possible routes for mycotoxin entry into human and animal foods*

1. Mould-damaged foodstuffs:	
(A) Agricultural produce,	e.g. cereals
	oilseeds (groundnuts)
	fruits
	vegetables
(B) Consumer foods (secondary infections)	
Compounded animal feeds (secondary infections)	
2. Residues in animal tissues and animal products,	
	e.g. milk
	dairy produce
	meat
3. Mould-ripened foods,	e.g. cheeses
	fermented meat products
	oriental fermentations
4. Fermenter products,	e.g. microbial proteins
	enzymes
	food additives, e.g. vitamins

Source: Adapted from Jarvis, 1976.

poultry and pigs, more susceptible to infections (Pier, Richard & Cysewski, 1980).

There is now an extensive literature which confirms the spasmodic occurrence of mycotoxins in human and animal food chains (WHO, 1979; Anon., 1980). The levels of contaminating mycotoxins vary from country to country, and within countries will be dependent on a wide range of environmental factors, agricultural practice and food hygiene. Many types of toxin producing fungi have been identified and over 200 fungal metabolites have been shown to have toxic potential to man and animals (Turner, 1971). However, most of these mycotoxins are primarily products of laboratory cultural manipulations and there is little evidence for wide natural occurrence.

In practice, only a relatively small number of mycotoxins can be found with any regularity in nature (WHO, 1979) and serious associated disease outbreaks are even less frequently documented. However, there is no denying the presence of low levels of mycotoxins in animal and human food chains and it is still a matter of debate whether such occurrences constitute a real long-term hazard to human and animal health.

Mycotoxin production by a given fungus has been shown to be dependent on three conditions:

    (a) the actual presence of the toxigenic fungus;

    (b) a suitable substrate for the growth of the fungus; and

    (c) an environment suitable for fungal growth.

Toxin production will only occur when all three conditions can be fulfilled. Each condition will involve many inter-relating factors which together or individually can affect mycotoxin formation (Hesseltine, 1976). Given the accepted presence of (a) and (b) the environment will undoubtedly be the main determining influence on mycotoxin production. Although most environmental parameters can influence mycotoxin production, moisture level of the substrate (a_w or water activity) and temperature have undoubted importance (Hesseltine, 1976; Lacey, Hill & Edwards, 1980).

There have been numerous studies attempting to quantify the types of toxigenic moulds in food and feed products (Mislevic, 1977). Although identification of toxigenic moulds may be of diagnostic value when outbreaks of mycotoxicoses occur, accurate conclusions can only be achieved by careful identification of the toxin(s) actually present in the suspected sample since:

    (a) the presence of the fungus is no assurance that it was actually producing the mycotoxin;

    (b) a given toxin may persist in a feedstuff when the fungus producing it is no longer present;

    (c) a given fungus may be capable of producing more than one toxin; and

    (d) a given toxin may be produced by different genera of fungi.

However, there is little doubt that the presence of a known toxigenic fungus in reasonable quantity does imply some degree of hazard.

Principles of sampling for fusaria and *Fusarium* toxins in field and stored products

The enumeration of fungi and mycotoxins in bulk organic materials and the interpretation of their significance can present many problems. Worthwhile enumeration will depend on (a) adequate sampling and subsampling so that the subsample examined is representative of the original bulk; (b) satisfactory storage for sampling until examination; (c) examination with the minimum possible delay to minimise changes in the microflora and mycotoxin production; and (d) use of appropriate enumerator techniques with suitable isolation media for fungal identification, together with sensitive and accurate methodologies for mycotoxin determinations (Lacey, Hill & Edwards, 1980).

The species of fungi found on grain can be classified into different ecological groups on the basis of their physiological requirements. The relative abundance of the fungi of these groups isolated from grain can be used to deduce the history of the sample and how well it may have been stored since harvest. Broadly fungi may be of two different types, viz. 'field' or 'storage'. Separation is not absolute and many fungi can show intermediate properties (Corry, 1978).

Field fungi grow on grain before harvest and generally do not develop further in storage. Such fungi will be favoured by wet weather during ripening. Many *Fusarium* spp. are distinctly field fungi. In contrast, storage fungi occur only in very small numbers before harvest and typically develop when the grain is transported and stored. The numbers and types of storage fungi depend on the conditions of storage, in particular the water activity of the grain and its temperature which can be increased by heating due to microbial respiration, as well as aeration, the use of chemical preservatives, the presence of foreign material, insect and mite infestation and finally the nutrient quality of the grain. The fusaria are generally not considered as storage fungi. However, some fusaria, viz. *F. culmorum*, *F. poae* and *F. tricinctum*, may sometimes grow during storage, e.g. in products with high water activity stored at low temperatures. The fungi isolated may well indicate the probable conditions of storage together with the possibility of mycotoxin formation (Lacey, Hill & Edwards, 1980).

The standard methods of enumerating grain fungi are by plating either whole grains (direct plating) or by serial dilutions of washings from the grains or suspensions of comminuted grain (dilution plating). Direct plating enables the proportions of grains carrying different fungi to be assessed. Several species can be detected on a single grain and this method often yields a wider range of species than dilution plating, particularly for fungi that do not produce abundant spores such as *Fusarium*. The level of penetration of the fungi can be determined by surface sterilisation of the grain with sodium hypochlorite and aseptic dissection of the grain and plating the individual fragments separately.

Dilution plating will favour those fungi which readily produce spores. Standardisation of techniques is essential, in particular the methods of preparing the primary suspension, the pipetting technique to be used in preparing serial dilutions, and plating. The methodology has been well-described by Flannigan (1977).

Choice of media can also be extremely critical. Potato dextrose agar is satisfactory for isolating fusaria while oxgall or OAES agars give good counts while restricting the growth of fast growing fungi such as *Rhizopus* (Lacey & Dutkiewicz, 1976). Most *Fusarium* spp. will grow well at 25–30 °C or room temperature and may be counted after six to seven days. As has been indicated in other chapters, the fusaria present their own unique problems with identification and there can be little doubt that the literature abounds with incorrect identification.

In any survey of fungi and mycotoxins in natural products such as cereal grains or animal feeds possibly the most important factor will be sampling technique. Moulds and mycotoxins are rarely uniformly distributed throughout organic material; rather their occurrence is spasmodic or uneven. Thus in large volumes of materials, e.g. cereal grains, great care must be taken to arrive at the situation where the final analysed samples are as truly representative of the whole mass as possible. Davis *et al.* (1980) have recently set out guidelines to be followed not only for achieving accurate sampling techniques but also for reducing the many problems that can occur between the time of sampling and actual analysis. Under inadequate handling conditions such as high humidity and temperature, mycotoxins may well be produced in the time period after withdrawal of sample when in fact the conditions of the original storage were less conducive to mould sporulation and mycotoxin formation.

Surveys on mould and mycotoxin levels in natural situations can be carried out on grain (or any other product) at any stage of production, harvesting, transport, processing or utilisation. Comparative surveys can be carried out in any part of the world and results compared if similar guidelines have previously been set out.

Each type of sampling will present different problems. If a lot, or consignment, has recently been blended by harvesting, loading and unloading, turning, grinding or any other means, a representative sample will be more easily obtained than if the lot had not been subjected to blending. Mould growth and toxin presence will invariably be in discrete regions and not uniform. Stream sampling, in which small samples are taken at regular intervals from a moving mass of material, is the most effective method of sampling. Probe sampling can only be truly effective if the bulk material has recently been mixed, otherwise this method will not produce representative samples. In order to obtain a good representative field sample from growing crops large numbers of widely distributed units or seeds must be gathered. Field samples are best taken at harvesting, thus ensuring wide selection and good random mixing.

In practice random sampling will achieve an understanding of the average distribution of moulds and toxins in agricultural or other materials. However, obviously moulded material should also be analysed as this may often be the material fed to animals and could be the cause of a mycotoxicosis. In particular, on-farm storage can achieve product imbalance due to leakage in storage bins causing local pockets of mould growth and possible toxin production. Such material can then leave the bin and be fed directly or compounded into feed for farm animals. Many on-farm outbreaks of a mycotoxicosis have arisen from such conditions.

General recommendations on sample size are difficult to make and will depend on actual particle size. Thus larger samples will be required with increasing particle size, e.g. corn as opposed to wheat as opposed to milled products. Larger samples will increase accuracy but the cost of product transportation and analysis will be restrictive. Cost factors will by necessity require to be reconciled with the accepted levels of accuracy. Whitaker & Dickens (1979) have determined the effect of sample size on sample distribution and lot distribution for mycotoxin distribution. In general, 10 lb samples have been shown to be sufficient for most survey purposes with shelled corn.

Since it is seldom feasible or even desirable to extract the entire sample for analysis, the large sample should be comminuted so that a subsample of the comminuted material can be extracted and analysed (Table 10.2). Larger subsamples are required for coarsely ground material than for finely ground material. Recommended methods for comminution and subsampling are given by AOAC (1975). In practice the original sample material should be ground to pass a No. 14 sieve, thoroughly blended and properly subdivided to a 1 kg sample. The entire 1 kg sample should then be ground to pass a No. 20 sieve, thoroughly blended and properly subdivided to 50 g analytical samples. Whenever possible the sample should be ground and subsampled immediately after collection.

Amplification of mould and mycotoxin concentration in samples after collection and before analysis has not been well studied. When subsamples cannot be immediately analysed they should be stored under refrigeration or dried.

Table 10.2. *Sampling schedules used for grain stored in bags or in bulk*

Bagged grain: Number of bags	Minimum number of sub-samples
< 4	Sample each bag
4–20	4
20–60	6
60–100	9
100–400	16
> 400	20
Bulk grain: Weight of grain	Minimum number of kg samples
< 100 kg	4
100 kg–1 t	6
1–3 t	10
3–5 t	12
5–25 t	20
>25 t	40

Source: From U.K. Fertilisers and Feeding Stuffs Regulations, 2973; Statutory Instrument No. 840, 1976.

Zearalenone

Zearalenone is the most studied of the *Fusarium* toxins primarily because of the well-documented oestrogenic effect of this compound on farm animals. It is said to be produced by strains of *Fusarium graminearum*, *F. tricinctum*, *F. oxysporum*, *F. sporotrichioides* and *F. moniliforme* (WHO, 1979), but may have a more limited distribution (see Chapter 12).

Zearalenone has been detected in hay, feed, corn, pig feed, sorghum, dairy rations and barley that had caused field problems in farm animals in England, Yugoslavia, Finland, Scotland, Japan, Africa and the United States (Table 10.3). Extensive reviews on the individual occurrences of zearalenone have been prepared by Hesseltine (1974), Stoloff (1976), Shotwell (1977), Bennett & Shotwell (1979) and Mirocha, Pathre & Christensen (1977).

The causal *Fusarium* spp. attack the developing seeds during periods of heavy rainfall and will proliferate on mature grains that have not dried because of wet weather at harvest or on grains that have been stored wet. In general low temperatures (12–14 °C) are needed to initiate and maintain high production of zearalenone.

Table 10.3. *Natural occurrence of zearalenone*

Commodity or Product	Examined because of	Country	Zearalenone (p.p.m.)
Hay	Infertility in dairy cattle	England	14.0
Feed	Infertility in cattle and swine	Finland	25.0
Corn	Hyperoestrogenism in farm animals	France	2.3
Animal feed	Hyperoestrogenism in cattle and swine	United States	0.1–2900
Corn		Yugoslavia	18
Corn	Poisoning in swine	Yugoslavia	2.5–35.6
Corn	Severe mould damage and swine refusal	Yugoslavia	0.7–14.5
Barley	Stillbirths, neonatal mortality, and small litters in swine	Scotland	0.5–0.75
Corn (freshly harvested	*Gibberella zeae* damage	United States	0.1–1.5
Corn (stored)	*A. flavus* damage	United States	ND – 92
Barley	Death in swine	Scotland	'Traces'
Feed	Field problems in animals	United States	Not stated
Grain sorghum	Head blight in sorghum	United States	Not stated
Corn	Swine hyperoestrogenism	Yugoslavia	35.6
Pig feed	Swine hyperoestrogenism	United States	50.0
Corn	Swine hyperoestrogenism	United States	2.7
Sorghum	Cattle abortion	United States	12.0
Corn	Swine abortion	United States	32.0
Silage		United States	87.3
Corn		England	306.0
Corn	Swine feed refusal	United States	2.5
Dairy ration	Cattle feed refusal, lethargy, anaemia	United States	1.0
Pig feed	Swine internal haemorrhaging	United States	0.1
Pig feed	Swine hyperoestrogenism	Yugoslavia	0.5
Pig feed	Swine infertility and abortion	United States	0.01

Source: From Shotwell, 1977.
Note: ND = Not detected.

Table 10.4. *Surveys of corn for zearalenone*

Year	Agency surveying	Origin of samples	Type of sample and source	Number of samples assayed	Per cent samples with indicated level of zearalenone (p.p.m.)			
					N.D.[1]	0.4	0.4–0.9	1.0–5.0
1967	N.R.R.C.	Corn belt	Grain inspection A.M.S.	283	99		1	
1968–69	N.R.R.C.	Export cargo	Grain inspection A.M.S.	293	98		2	
1972	F.D.A.	Corn belt[2]	Elevator and food processing	223	83	9	4	4
1973	F.D.A.	Corn belt	Farm and country elevator	169	90	10		
1973	F.D.A.	South[3]	Farm and country elevator	146	99	1		
	University of Minnesota	Mexico	For human consumption	139[4]	96			

Source: From Shotwell, 1977.
Note:
[1] N.D. = not detected.
[2] Area where potential for *Fusarium* contamination was considered to be high or where *Fusarium* damage had been reported.
[3] Includes Southeast, Appalachia, Southeast MO, KY, TN, OK, TX, and CA.
[4] Levels in six positive samples were not reported.

Table 10.5. *Natural occurrence of zearalenone in feeds associated with hyperoestrogenism in swine*

Feed sample	Level of contamination (mg/kg)
Maize kernels (Minnesota)	0.1–0.15
Dry sow ration (Vancouver)	0.15
Farrowing ration (Vancouver)	0.068
Dry sow ration (Vancouver)	0.15
Corn kernels (Vancouver)	0.20
Dry sow ration (Vancouver)	0.25
Lactation ration (Vancouver)	1.00
Gestation ration (Vancouver)	0.5
Milo (Minnesota)	2.5–5.6
Sesame meal (Univ. of Minnesota)	1.5
Corn kernels (Ohio)	0.12
Mixed feed corn (Ohio)	0.12
Corn kernels (Minnesota)	6.4
Commercial pelleted mixed feed (Minnesota)	6.8

Source: From Mirocha, Pathre & Christensen, 1977.

Corn (maize) and barley are the two main cereal crops that regularly support *Fusarium* growth and toxin formation. The presence of zearalenone in cereal grains has been mainly established either as a result of investigations of field outbreaks of mycotoxicosis, or from surveys of grain collected at dispersal points in the marketing system.

Surveys have been made of corn, corn products, wheat, soybeans, and grain sorghum moving in commercial channels in the U.S.A. In general, levels were higher in corn when there had been unusually wet harvesting conditions (Table 10.4). Soybeans were remarkably free of zearalenone in tested samples. In many cases investigations were initiated because of obvious mould damage to the grains. Infected kernels appear bleached and in severe infections can possess reddish discoloration and are chaffy.

Zearalenone has regularly been found in a number of mixed feeds implicated in hyperoestrogenism in pigs and cattle (Table 10.5), and in most cases has been due to the use of contaminated corn or barley. There is growing evidence that if zearalenone is detected in corn there is every possibility that other *Fusarium* toxins will also be present (Thiel, Marasas & Meyer, 1982). The co-existence of zearalenone with other non-*Fusarium* toxins has also been documented (Table 10.6).

Table 10.6 *Co-existence of zearalenone with other mycotoxins*

Commodity	Level of zearalenone (p.p.m.)	Other mycotoxin	Level of mycotoxin (p.p.m.)
Corn	1.2	Aflatoxin	0.037
Corn	0.6	Aflatoxin	0.006
Corn	14.5	Ochratoxin	3.1
Corn[1]	N.D.–92	Aflatoxin	N.D.–1.7
Corn[2]	0.2–10.4	Aflatoxin	0.1
Barley	Not stated	T-2 toxin	Not stated

Source: From Shotwell, 1977.
Note: [1] Samples taken from various parts of bin and analysed separately. Results are not representative of entire lot.
[2] Corn that was obviously mould damaged.

The presence of zearalenone in maize beer in Africa has been well-documented. Home-brewed and commercial beers have been shown to have zearalenone levels as high as 4.6 mg/litre. Zearalenone has also been detected in corn mash used in fermentations, in sour drinks, sour porridge and local beers prepared by fermenting corn or sorghum meals (WHO, 1979).

Trichothecenes

The trichothecene toxins are a group of metabolites character-ised by a tetracyclic 12,13-epoxy-trichothec-9-ene skeleton. There are over 40 naturally occurring derivatives produced predominantly by species of *Fusarium* but also by some species of *Myrothecium*, *Stachybotrys* and *Verticimonosporium* (Table 10.7). The trichothecene deriva-tives most commonly found in cereal grains and feed stuffs are T-2 toxin, diacetoxyscirpenol (DAS), deoxynivalenol (vomitoxin) and nivalenol (NIV). One or more of the trichothecenes have been isolated from strains of *Fusarium* species referred to as *F. episphaeria*, *F. lateritium*, *F. nivale*, *F. oxysporum*, *F. rigidiusculum*, *F. solani*, *F. roseum*, *F. tricinctum*, *F. sporotrichioides* (Table 10.7).

The natural occurrence of trichothecenes in cereal grains and feeds is shown in Tables 10.8 and 10.9. Vomitoxin has been isolated from corn and mixed feed more frequently than the other derivatives reported and moreover it is frequently found together with the fungal oestrogen, zearalenone. Vomitoxin has frequently been associated with corn

Table 10.7. *Trichothecene-producing fungi*

	Trichothecene type[1]		
A	B	A and B[2]	C
Fusarium tricinctum	*F. nivale*	*F. equiseti*	*Myrothecium verrucaria*
F. roseum	*F. episphaeria*	*F. scirpi*	*Myrothecium roridum*
F. roseum 'culmorum'	*F. roseum*	*F. oxysporum*	*Stachybotrys atra*
F. roseum 'avenaceum'		*F. sp.* K50 10	*Verticimonosporium*
F. roseum 'scirpi'			*diffractum*
F. sporotrichioides			
F. poae			
F. solani			
F. rigidiusculum			
F. lateritium			
F. semitectum			
F. equiseti			

Source: From Ueno, 1977.
Note: [1] Type A have H,OH or H,H at position 8 e.g. T-2 toxin, HT-2 toxin, diacetoxyscirpenol and neosolaniol; type B have a ketone group at position 8 e.g. nivalenol, monoacetylnivalenol, diacetylnivalenol and deoxynivalenol; type C have a macrocyclic constituent joining positions 4 and 15 e.g. the roridins, verrucarins, satratoxins and vertisporin. See Fig. 12.8 for the numbering system of the trichothecene molecule.
[2] This group includes diacetoxyscirpenol, diacetylnivalenol 7-hydroxydiacetoxy-scirpenol and 7,8-dihydroxydiacetoxyscirpenol.

Table 10.8. *Natural occurrence of vomitoxin on cereal grains and feed*

Country	Commodity	Concentration (μg/g)
U.S.A.	Corn	40
U.S.A.	Corn	15, 18, 20, 28
U.S.A.	Corn	28
U.S.A.	Corn	0.7
U.S.A.	Corn	12
U.S.A.	Corn	1, 1.8
U.S.A.	Commercial, mixed feed	0.04–0.06
U.S.A.	Mixed feed	0.4, 0.6, 1
Japan	Barley	7.3
France	Corn	0.1–0.6
Canada	Corn	7.9
Canada	Feed	1.2
Austria	Corn	1.3, 7.9
South Africa	Corn	2.5
Zambia	Corn	7.4

Source: From Vesonder & Hesseltine, 1981.

Table 10.9. *Natural occurrence of trichothecene toxins in feedstuff*

Sample no.		Concentration (μg/kg)	Diagnosis	Feedstuff
FS-382	Diacetoxyscirpenol	500	Haemorrhagic bowel syndrome in swine	Mixed feed (Univ. Minn.)
FS-404	Diacetoxyscirpenol	380	Haemorrhagic bowel syndrome in swine	Mixed feed (Univ. Minn.)
FS-356	Deoxynivalenol	1800	Feed refused by swine	Maize kernels (Michigan)
FS-362	Deoxynivalenol	1000	Feed refused by swine	Maize kernels (Indiana)
FS-398A	Deoxynivalenol	100	Feed refused by swine	Maize kernels (Ohio)
FS-63	Deoxynivalenol	40–60	Feed refused by swine and bloody stools	Commercial pelleted mixed feed
FS-417	T-2 toxin	76	Bloody stools, bovine	Mixed feed (Nebraska)
FS-483	Deoxynivalenol	1000	Vomiting in dogs	Mixed feed (Iowa)
FS-489	Deoxynivalenol	1000	Feed refused by swine.	Mixed feed (Minnesota)
FS-419A	Deoxynivalenol	2500	—	Maize kernels (S. Africa)
FS-516B	Deoxynivalenol	7400	—	Maize kernels (Zambia)
FS-543B	Deoxynivalenol	75	Cows – C.N.S. irritations, irregular heat cycles	Maize (Nebraska)
FS-570A	Deoxynivalenol	25	Feed refused by swine	Ground corn (New York)
FS-570C	Deoxynivalenol	200	Dogs – refusal, emesis	Dry extruded dog food (NY)
FS-570D	Deoxynivalenol	120	Feed refusal by swine	Ground corn (New York)
FS-570E	Deoxynivalenol	550	Feed refused by swine	Maize kernels (New York)

Source: From Pathre & Mirocha, 1979.

Table 10.10. *Incidence of* Fusarium *species and natural occurence of* Fusarium *toxins in Transkeian maize*

		Low standard samples A samples (36)			High standard samples B samples (36)		
		Mean	Range	% positive	Mean	Range	% Positive
F. moniliforme	% kernels infected	13.5	0–61	86	20.1	0–92	89
F. moniliforme var. *subglutinans*	"	24.5	0–78	92	40.8	1–94	100
F. graminearum	"	8.8	0–36	92	31.9	0–86	89
Total *Fusarium*	"	47.0	1–135	100	93.5	55–140	100
Moniliformin	mg/kg	2.1	0–12	42	10.9	0–45	92
Zearalenone	"	—	0–0.02	3	1.0	0.02–5.36	100
Deoxynivalenol	"	0.04	0–0.82	39	1.0	0–15.8	81
Nivalenol	"	0.02	0–0.24	17	0.3	0–1.41	39

Source: From Thiel, Marasas & Meyer, 1982.

Table 10.11. *Correlations between the incidence of* Fusarium *species and the concentration of* Fusarium *toxins in Transkeian maize (B samples)*

Incidence of *Fusarium* species	Mycotoxin content			
	Moniliformin	Zearalenone	Deoxynivalenol	Nivalenol
F. moniliforme	$R = 0.108$	$R = -0.430$	$R = -0.013$	$R = -0.361$
	N.S.	$P < 0.01$	N.S.	$P < 0.05$
F. moniliforme	$r = 0.611$	$r = -0.268$	$r = -0.031$	$r = -0.434$
var. *subglutinans*	$p < 0.001$	N.S.	N.S.	$p < 0.01$
F. graminearum	$r = -0.557$	$r = 0.720$	$r = 0.158$	$r = 0.823$
	$p < 0.001$	$p < 0.001$	N.S.	$p < 0.001$
Total *Fusarium*	$r = 0.011$	$r = 0.238$	$r = 0.096$	$r = 0.355$
	N.S.	N.S.	N.S.	$p = 0.05$

Source: From Thiel, Marasas & Meyer, 1982.

refused by swine, and vomiting in dogs and cows in South Africa (Pathre & Mirocha, 1979).

Although the trichothecenes have only been found very sporadically in natural products this may well be due to lack of extensive examination. There is little doubt that species of *Fusarium* are widely distributed geographically and associated toxin formation must be expected under specific conditions. Recent chemical and mycological data from the Transkei with maize indicated a very high rate of contamination by *Fusarium* species and the toxins they produce (Thiel, Marasas & Meyer, 1982). As expected, samples of maize intended for beer brewing or animal feeding (generally mouldy) showed a much higher rate of contamination as well as a much higher concentration of toxins than the samples from maize intended for human consumption (generally mould-free). The percentage of kernels infected with *F. graminearum* was highly significantly correlated with zearalenone ($r = 0.72$; $p < 0.001$) (Table 10.10). All 43 toxic isolates of *F. graminearum* produced zearalenone.

The detailed studies of the Transkei have also shown significant negative correlation between the incidence of *F. moniliforme* and the levels of zearalenone and nivalenol, between *F. moniliforme* var. *subglutinans* and nivalenol, and also between *F. graminearum* and moniliformin (Table 10.11). This implies that samples having high levels of infection by *F. moniliforme* and *F. moniliforme* var. *subglutinans* and containing moniliformin usually had low levels of infection by *F. graminearum* and *vice versa*. The relationship between diacetoxyscirpenol and T-2 toxin and the other toxins is being further studied by these workers.

REFERENCES

Anon. (1980). Survey of mycotoxins in the United Kingdom. *Food Surveillance Paper No. 4*. HMSO, London.

AOAC (1975). *Official Methods of Analysis. 12th Ed. AOAC*, Chapter 26. USA: Arlington.

Bennett, G. A. & Shotwell, O. L. (1979). Zearalenone in cereal grains. *Journal of the American Chemical Society*, **56**, 812–19.

Corry, J. E. L. (1978). Relationships of water activity to fungal growth. In *Food and Beverage Mycology*, ed. L. R. Beuchat, pp. 45–82. Connecticut: Avi Publishing Company.

Davis, W. D., Dickens, J. W., Freie, R. L., Hamilton, P. B., Shotwell, O. L., Wyllie, T. D. & Fulkerson, J. F. (1980). Protocols for surveys, sampling, post-collection handling and analysis of grain samples involved in mycotoxin problems. *Journal of the Association of Official Analytical Chemists*, **63**, 95–102.

Emmons, C. W., Binford, C. H., Utz, J. P. & Kwon-Chung, K. J. (1977). *Medical Mycology*. Philadelphia: Lea & Febiger.

Flannigan, B. (1977). Enumeration of fungi and assay for ability to degrade structural and storage components of grain. In *Biodeterioration Investigation Techniques*, ed. H. Walters, pp. 185–99. London: Applied Science Publishers.

Hesseltine, C. W. (1974). Natural occurrence of mycotoxins in cereals. *Mycopathologia et Mycologia Applicata*, **53**, 141–53.

Hesseltine, C. W. (1976). Conditions leading to mycotoxin contamination of foods and feeds. In *Mycotoxins and other Fungal Related Food Problems*, ed. J. V. Rodricks, pp. 1–22. Washington: American Chemical Society.

Jarvis, B. (1976). Mycotoxins in food. In *Microbiology in Agriculture, Fisheries and Food*, ed. F. A. Skinner and J. G. Carr, pp. 251–67. London: Academic Press.

Lacey, J. & Dutkiewicz, J. (1976). Methods for examining the microflora of moulding hay. *Journal of Applied Bacteriology*, **41**, 13–27.

Lacey, J., Hill, S. T. & Edwards, M. A. (1980). Microorganisms in stored grains: their enumeration and significance. *Tropical Stored Products Information*, **39**, 19–32.

Mirocha, C. J., Pathre, S. V. & Christensen, C. H. (1977). Zearalenone. In *Mycotoxins in Human and Animal Health*, ed. J. V. Rodricks, C. W. Hesseltine and M. A. Mehlman, pp. 345–64. Illinois: Pathotox Publishers Inc.

Mislevic, P. B. (1977). Toxigenic fungi in foods. In *Mycotoxins in Human and Animal Health*, ed. J. V. Rodricks, C. W. Hesseltine and M. A. Mehlman, pp. 469–78. Illinois: Pathotox Publishers Inc.

Moss, M. O. (1977). Aspergillus mycotoxins. In *Genetics and Physiology of Aspergillus*, ed. J. E. Smith and J. A. Pateman, pp. 499–525. London: Academic Press.

Pathre, S. V. & Mirocha, C. J. (1979). Trichothecenes: natural occurrence and potential hazard. *Journal of the American Oil Chemists' Society*, **56**, 820–3.

Pier, A. C., Richard, J. L. & Cysewski, S. J. (1980). Implications of mycotoxins in animal disease. *Journal American Veterinary Medical Association*, **176**, 719–24.

Shotwell, O. L. (1977). Assay methods for zearalenone and its natural occurrence. In *Mycotoxins in Human and Animal Health*, ed. J. V. Rodericks, C. W. Hesseltine and M. A. Mehlman, pp. 403–13. Illinois: Pathotox Publishers Inc.

Stoloff, L. (1976). Occurrence of mycotoxins in foods and feeds. In *Mycotoxins and Other Fungal Related Food Problems*, ed. J. V. Rodericks, pp. 23–50. Washington: American Chemical Society.

Thiel, P. G., Marasas, W. F. O. & Meyer, C. J. (1982). Natural occurrence of *Fusarium* toxins in maize from Transkei. In *Mycotoxins and Phycotoxins*, pp. 126–9. V IUPAC Symposium, Vienna.

Turner, W. B. (1971). *Fungal Metabolites*. London: Academic Press.

Ueno, Y. (1977). Trichothecenes: overview address. In *Mycoxins in Human and Animal Health*, ed. J. V. Rodericks, C. W. Hesseltine & M. A. Melhman, pp. 189–207. Illinois: Pathotox Publishers Inc.

Vesonder, R. F. & Hesseltine, C. W. (1981). Vomitoxin: natural occurrence on cereal grains and significance as a refusal and emetic factor to swine. *Process Biochemistry* (1980/81) Dec./Jan. 12–16.

Whitaker, T. B. & Dickens, J. (1979). Estimation of the distribution of lots of shelled peanuts according to aflatoxin concentrations. Quoted in Davis *et al.* (1980).

WHO (1979). *Environmental Health Criteria*. 11. *Mycotoxins*. World Health Organisation, United Nations Environment Programme, Geneva.

11
The detection and analysis of *Fusarium* mycotoxins*

J.GILBERT

Ministry of Agriculture, Fisheries and Food, Food Laboratory (Norwich), Haldin House, Old Bank of England Court, Queen Street, Norwich, NR2 4SX, U.K.

Introduction

Species of *Fusarium* are capable of simultaneously producing more than one toxin as secondary metabolites of which principally zearalenone together with the trichothecenes are recognised as being the most important. The detection and analysis of these mycotoxins in biological materials involves extraction and invariably an extensive chemical clean-up to remove potential interferences prior to separation by thin-layer, high performance or gas–liquid chromatography (GC), or a combination of these methods. Detection is ultimately by colorimetric or fluorimetric visualisation, by conventional GC detection or by mass spectrometry (GC–MS). This chapter critically reviews the relative merits of these possible chemical and instrumental techniques with regard to ease of analysis, sensitivity, specificity and quantitative precision.

The range of the problem

A diversity of analytical techniques ranging from the simplicity of thin-layer chromatography (t.l.c.) to the sophistication of combined gas chromatography–mass spectrometry (GC–MS) have been applied to the analysis of *Fusarium* mycotoxins. Whichever of these methods is employed and whether the toxin to be analysed is either zearalenone, one of the 75 known trichothecenes or another metabolite such as moniliformin, an extraction procedure followed by extensive chemical clean-up and concentration of the toxin in a final extract is essential. The extraction must be reproducible, as near to quantitative as possible and must have been evaluated on naturally contaminated materials. This is important to ensure efficient removal of the toxins as there might otherwise be binding to the matrix not easily simulated by spiking experiments. The extent of subsequent clean-up is then dictated by the specificity of ultimate detection, but it must be extensive if one is employing say t.l.c. and although it can be less so using a highly specific detection technique like mass spectrometry, even in the latter case large amounts of extraneous material in the final extracts are undesirable.

Of the *Fusarium* toxins, zearalenone is probably the simplest to analyse because the fluorescent properties of this compound make it amenable to sensitive detection after t.l.c. or high performance liquid chromatography (h.p.l.c.). In contrast the trichothecenes are particularly difficult to analyse as they have no native fluorescence so that t.l.c. detection must involve non-specific colorimetric visualisation which will also inevitably be less sensitive than fluorescence. Additionally the trichothecenes comprise a large group of compounds which, although possessing common structural features including the tricyclic ring system and the 12,13-epoxide group, also contain many variations in both position and type of substituent. This means that at the outset a decision is required as to which specific trichothecenes within the group one wishes to assay. Although supposedly 'general' trichothecene visualisation procedures for t.l.c. or post-column derivatisation h.p.l.c. are available, based on reaction at the epoxide group, there are as yet no means of making a total estimation of trichothecene levels.

Trichothecenes can be made sufficiently volatile for GC analysis by derivatisation of the free hydroxy groups (usually by formation of the trimethylsilyl (TMS) or heptafluorobutyryl (HFB) derivatives) and can be detected by flame ionisation (f.i.d.) or electron capture (e.c.d.). Although the latter is preferable to f.i.d., being both more specific and sensitive, the complexity of many extracts from cereals and feeding stuffs still means that erroneous positive results can be fairly common. It is therefore desirable to confirm positive results by GC–MS using selected ion monitoring, whence by a choice of suitable ions a number of trichothecenes can be simultaneously assayed during the course of a single GC run.

Although, of the *Fusarium* mycotoxins, the trichothecenes and zearalenone have received the most attention, there are also the possibilities of other metabolites such as moniliformin being naturally present, and the analysis of these compounds is also discussed in this chapter.

Analysis of trichothecenes

More than 75 structurally different trichothecenes have been identified, although the most frequent reports of natural occurrence concern a limited number including T-2 toxin, HT-2 toxin, diacetoxy-scirpenol, neosolaniol, Fusarenon-X, nivalenol and deoxynivalenol (structures shown in Fig. 11.1). Not all of this small group are available commercially as reference standards and method development, and thereby the obtaining of surveillance data, has been somewhat restricted. For the detection and analysis of trichothecenes, especially for quantification, an initial decision is required as to which individual compounds are to be monitored.

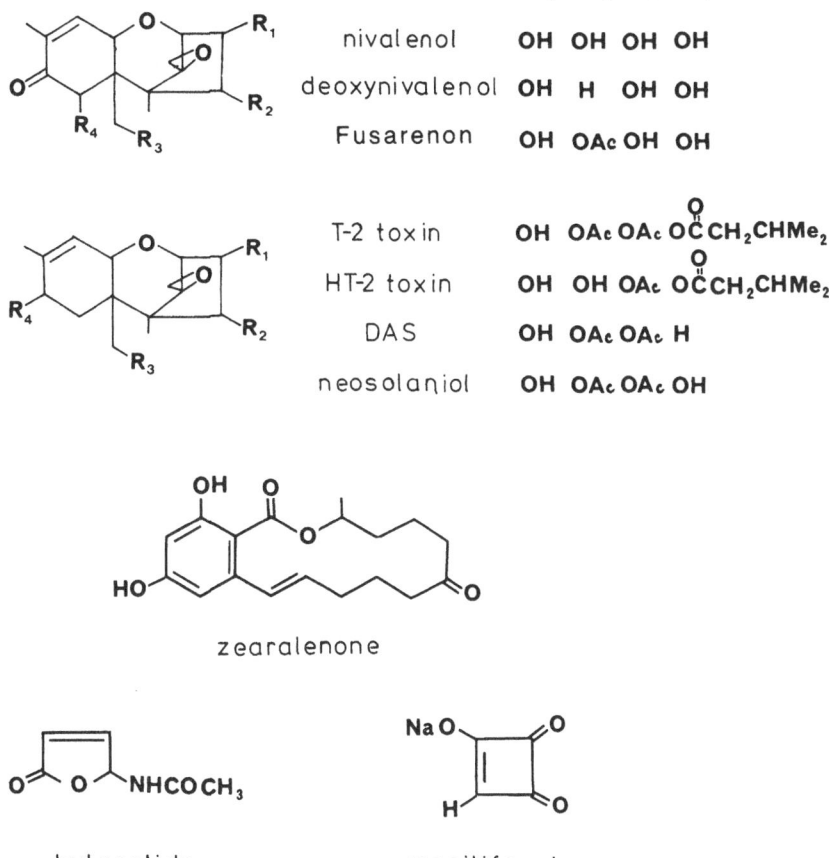

Fig. 11.1. Structures of *Fusarium* mycotoxins.

Extraction and clean-up

A number of workers have classified the trichothecenes into groups on the basis of extractability into solvents of differing polarity (e.g. Pathre & Mirocha, 1977). For example, T-2 toxin, HT-2 toxin, neosolaniol and diacetoxyscirpinol are very soluble in ethyl acetate, acetone, chloroform and dichloromethane whilst the highly hydroxylated trichothecenes nivalenol, deoxynivalenol and Fusarenon-X are soluble in very polar solvents like aqueous methanol or aqueous acetonitrile. For most practical purposes it is desirable to choose a solvent capable of extracting all the trichothecenes simultaneously from both groups; to this end aqueous methanol has been most frequently employed.

Initial extracts from cereals and feedingstuffs usually contain proteins, lipids and pigmented substances, all of which interfere with subsequent analysis. There are unfortunately no rapid clean-up procedures and most methods involve a number of stages often including protein precipitation with, for example, ammonium sulphate (Scott, Lau & Kanhere, 1981), and filtration or centrifugation of the extract followed by liquid/liquid partition (Scott *et al.*, 1981; Forsyth *et al.*, 1977). This partitioning can be carried out with success into ethyl acetate and the solvent then removed prior to further chromatographic clean-up. A silica gel column purification (or a Sep-Pak cartridge) can be employed eluting with dichloromethane/methanol (Scott *et al.*, 1981), although Amberlite XAD resin chromatography (Kamimura *et al.*, 1981), Sephadex LH-20 size-exclusion chromatography (Vesonder *et al.*, 1976), h.p.l.c. (Bennet *et al.*, 1981) and preparative t.l.c. (Ilus, Niku-Paavola & Enari, 1981) have all been employed with equal success.

Thin layer chromatography

The separation of standard mixtures of trichothecenes by silica gel t.l.c. presents no difficulty and many solvent systems have been reported, usually utilising mixtures of chloroform and methanol. Combinations of benzene/acetone, ethyl acetate/hexane, ethyl acetate/toluene and benzene/tetrahydrofuran have also been employed successfully. The main problem is in the detection of trichothecenes after t.l.c. because, unlike zearalenone and aflatoxins, they exhibit no fluorescent properties nor do they absorb appreciably in the ultraviolet. The use of a visualisation procedure is therefore necessary, and a summary of the published methods is given in Table 11.1. Of these methods sulphuric

Table 11.1. *TLC visualisation procedures for trichothecenes*

Procedure	Trichothecene	Limit of detection* (μg per spot)	Colour	Reference
p-Anisaldehyde (MeOH, acetic acid, H_2SO_4 soln.)	Deoxynivalenol	0.05	Yellow	Ilus et al. (1981).
	Diacetoxyscirpenol	0.10	Purple	Scott et al. (1970).
	T-2 toxin	0.10	Brown	
	HT-2 toxin	0.20	Brown	
20% H_2SO_4 soln.	Deoxynivalenol	0.05	Yellow	Ueno et al. (1973).
	Diacetoxyscirpenol	0.20	Purple	
	T-2 toxin	0.20	Grey	
	HT-2 toxin	0.50	Grey	
10% Aluminium chloride	Deoxynivalenol	0.10	Blue (fluorescent)	Kamimura et al. (1981).
	Nivalenol	0.10		
	Fusareron-X	0.10		
4-(p-nitrobenzyl) pyridine	All trichothecenes	0.02–0.2	Blue spots	Takitani et al. (1979).
Nicotinamide/2-acetylpyridine	All trichothecenes	0.02–0.05	Light blue (fluorescent)	Sano et al. (1982).

*Determined as pure reference standards.

acid charring (Ueno *et al.*, 1973) is probably the least specific, although when employed under the prescribed conditions the developed colour is to some extent characteristic of the individual trichothecene, varying from grey through purples and browns to yellow. Similarly p-anisaldehyde (Scott, Lawrence & van Walbeek, 1970) gives characteristic colours for different trichothecenes, whilst aluminium chloride (Kamimura *et al.*, 1981) gives blue fluorescent spots for the 8-ketotrichothecenes only, such as nivalenol and deoxynivalenol, with a good sensitivity. However, both these reagents lack structural selectivity, and there is always a high risk of erroneous assignment of the presence of trichothecenes to other components present in the final cleaned-up extracts, especially from complex matrices such as cereals.

Two other visualisation procedures reviewed in Table 11.1 were developed by making use of reactions at the 12,13-epoxy group in the trichothecene nucleus. 4-(p-nitrobenzyl)pyridine is sprayed onto the plate, which is heated for 30 minutes at 150 ° followed by spraying with tetraethylene pentamine (Takitani *et al.*, 1979). The trichothecenes can be seen as blue spots on a light blue background and the reported sensitivity is adequate to achieve a limit of around 50 μg/kg using a reasonably extensive clean-up procedure. The other epoxide-specific procedure utilising nicotinamide and 2-acetylpyridine produces fluorescent spots for trichothecenes at an even better sensitivity (Sano *et al.*, 1982). However, this method is both elaborate and time-consuming involving spraying the developed plate with nicotinamide solution, heating at 160 ° for 15 minutes, spraying with 2-acetylpyridine solution followed by potassium hydroxide and then allowing to react for 30 minutes at room temperature. After dipping in formic acid solution and evaporating the solvent, the plate is finally heated for 4 minutes at 100 ° and the trichothecenes can be seen as light blue fluorescent spots on a weakly fluorescent dark blue background.

In particular when employing t.l.c. care must be taken to guard against interferences in the extract causing erroneous assignment of identity. The use of 2-dimensional t.l.c. improves the separation of trichothecenes from other components as does the procedure of successive developments of the plate in different solvent systems, and both these techniques are strongly recommended for examining biological extracts. Additionally the use of more than one visualisation procedure improves the confidence in identification and, for example, a component of the correct R_f in 2-dimensional t.l.c. for deoxynivalenol which shows blue fluorescence when sprayed with aluminium chloride, is yellow after spraying with *p*-anisaldehyde and shows the characteristic blue general trichothecene reaction with 4-(*p*-nitrobenzyl)pyridine has a high probability of, in fact, being deoxynivalenol.

If quantification of trichothecenes is attempted by t.l.c. it should be remembered that fluorescence is often quenched by impurities in the sample. Hence the reference trichothecenes for calibration need to be mixed with control sample extract, as a comparison with the pure toxin alone can often be misleading. Visual quantification of the intensities of the fluorescent spots is possible with the normal limitations of precision although with densitometry this can be much improved (Sano *et al.*, 1982). From the limits of detection given in Table 11.1 for the pure trichothecenes (i.e. from 0.02 – 0.5 μg) calculated on the basis of using an average clean-up procedure which produces 10 g equivalents from say a cereal in a final extract of 100 μl, of which 10 μl is applied to a t.l.c. plate, this gives a sensitivity of the method ranging from 20 to 500 μg/kg. Obviously in some instances more than one-tenth of the final extract could be applied to the t.l.c. plate or with a more elaborate procedure it may be possible to utilise say 50 g of cereal. Nevertheless this simple calculation is based on an ideal situation of the pure toxins visualised against a clean background, and the reality of an extract with fluorescence quenching and other coloured interferences means that it is probably optimistic to expect t.l.c. procedures to be reliable for screening purposes below a level of contamination of around 100 μg/kg on average, and at a level considerably higher for some of the more unfavourable toxins, particularly when using the less sensitive visualisation procedures.

Gas chromatography

After derivatisation of the free hydroxy groups in the trichothecenes to form trimethylsilyl (TMS) ethers or heptafluorobutyryl (HFB) esters, they become sufficiently volatile for gas chromatographic analysis. Derivatisation is easily carried out by treating the dry extract with bis(trimethylsilyl) trifluoroacetamide (BSTFA), evaporating to dryness and re-dissolving in a suitable volatile solvent, or alternatively by treating with a slightly more powerful silylating reagent, e.g. TRI-SIL 'TBT' which, having a higher boiling point, is more difficult to remove but can be directly injected for GC analysis. Formation of the heptafluorobutyryl derivative involves a slightly more elaborate procedure of heating the extract with heptafluorobutyryl-imidazole for 1 hour at 60 °, washing the extract with sodium bicarbonate solution and dis-

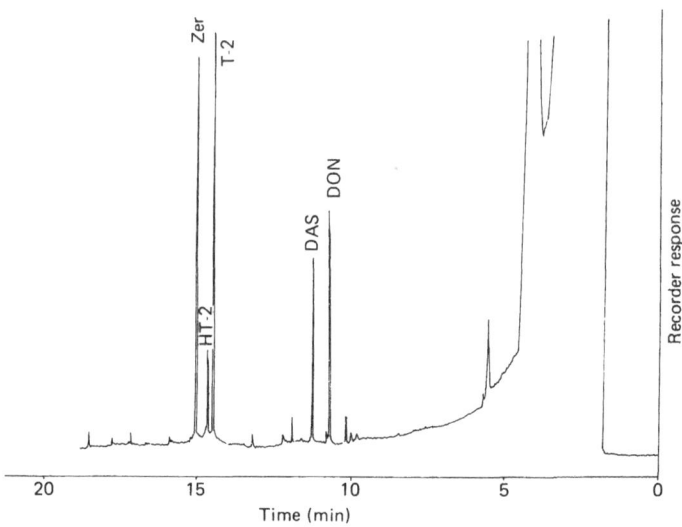

Fig. 11.2. Chromatogram illustrating separation of a mixture of TMS trichothecene reference standards. DON, deoxynivalenol; DAS, diacetoxyscirpenol; T-2, T-2 toxin, HT-2, HT-2 toxin; Zer, Zearalenone. 25 m × 0.3 mm glass capillary column coated with OVl operated at 90 °C for 2 min programmed at 30 °/min to 250 °C and then at 8 °/min to 300 °C. Splitless injection at 240 °C. FID detection.

solving in hexane prior to GC analysis. Once derivatised the trichothecenes can be analysed on non-polar stationary phases by either packed, or, preferably, capillary column GC. A typical chromatogram for a derivatised mixture of reference standards is shown in Fig. 11.2.

Normal routine choice of detection in GC is by flame ionisation which is far more sensitive for trichothecenes than t.l.c., having a limit of detection of around 1 ng per component on-column. Although the high resolution of capillary columns means that this technique is intrinsically more selective than t.l.c., it is nevertheless utilising a non-specific detection method. This means that with the complexity of a biological extract one can easily make a wrong assignment of the identity of a peak purely on the basis of retention times. An illustration of this complexity can be seen from Fig. 11.3, which is a chromatogram from a rice medium which was inoculated with *Fusarium nivale* and stored under optimum conditions for trichothecene production. Although the major components in this cleaned-up extract are in fact the trichothecenes nivalenol and Fusarenon-X it would be quite easy on retention times alone to wrongly assign many of the other peaks in the chromatogram.

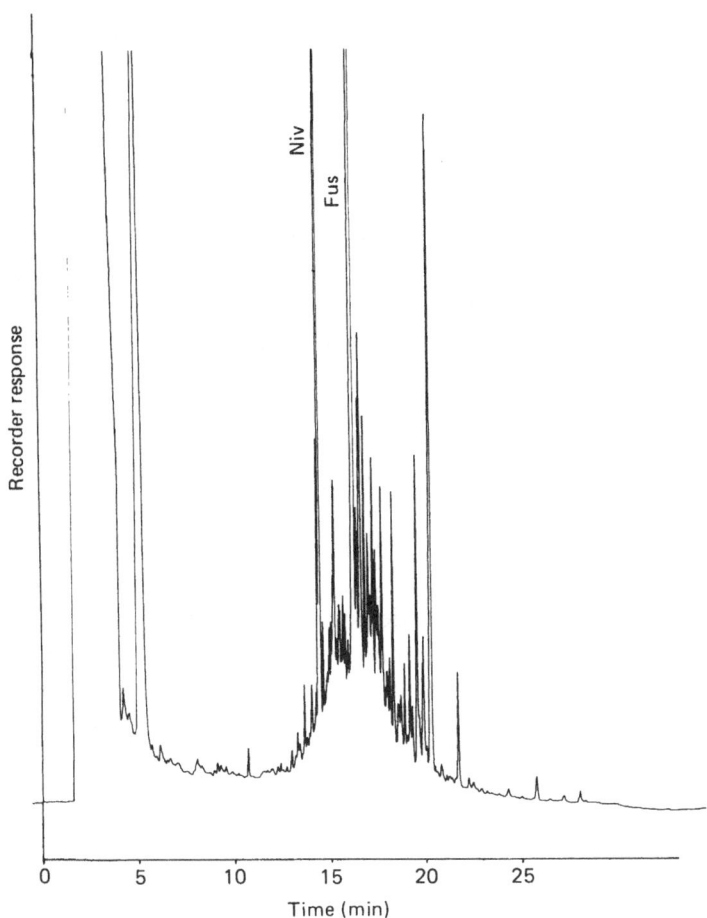

Fig. 11.3. Chromatogram of rice extract inoculated with *Fusarium nivale*. TMS derivatised trichothecenes: Niv, nivalenol; Fus, Fusarenon-X. GC conditions using FID as for Fig. 11.1. but programmed from 145 °C (isothermal for 2 min) to 270 °C at 15 °/min.

The trichothecene TMS derivatives are in themselves electron capturing but only offer a marginally improved sensitivity over the f.i.d. If one is proposing to carry out electron-capture GC the heptafluorobutyrate derivative offers the greater sensitivity, and this is the best choice for an inexpensive screening procedure.

In the author's experience, however, background 'chemical' interference generally makes for uncertainties in detection below 50 μg/kg levels using e.c.d., and also in a number of cereals false positives have occurred due to co-chromatographing interfering compounds from the commodities. For unequivocal confirmation mass spectrometry is essential, but as this sophisticated instrumentation is not freely available, the degree of confidence in a result can be increased by utilising a combination of other assay procedures. For example, one might proceed initially by checking for correct R_f and colour changes by t.l.c. for a suspected trichothecene in a cereal. This could be followed by scraping the spot from the plate, derivatising and analysing by G.C. with electron capture detection on say two different stationary phases. Agreement on the approximate quantitative levels by both of these differing chromatographic techniques would provide good substantiating evidence of identification.

Combined gas chromatography–mass spectrometry (GC–MS)

Unequivocal confirmation of identification by GC–MS involves obtaining a full spectrum and this requires approximately 10 to 20 ng per trichothecene injected on-column, which equates with a contamination level of about 100 to 200 μg/kg for an average clean-up procedure. Full spectra of both a TMS deoxynivalenol reference standard and a component identified in contaminated maize are illustrated in Fig. 11.4. Good agreement, in this type of example, would be expected both qualitatively for the presence of all the major ions as well as quantitatively – i.e. with respect to their relative abundancies. Additional ions in the maize extract might indicate co-chromatographing contaminants, and careful checking of the spectrum against the reference standard is therefore essential. It is also worth noting that the reference spectrum should be obtained under identical mass spectrometer operating conditions to that of the unknown, as small changes in certain parameters, e.g. source temperature, can seriously alter the relative abundance of the ions in the spectrum.

The TMS trichothecenes do not show particularly characteristic spectra under electron impact conditions either giving a low intensity molecular ion or no molecular ion at all, and the predominant fragmentation pattern is one of uncharacteristic low mass ions associated with the silylated portion of the molecule. For this reason chemical ionisation is a very useful ancillary mass spectrometric technique, whereby the introduction of a reagent gas into the source of the mass spectrometer gives a modified fragmentation pattern with a high proportion of ion-current associated with an even-electron pseudo-molecular ion. In Fig. 11.4 the chemical ionisation spectrum, using methane as the reagent gas, for deoxynivalenol is shown with an intense

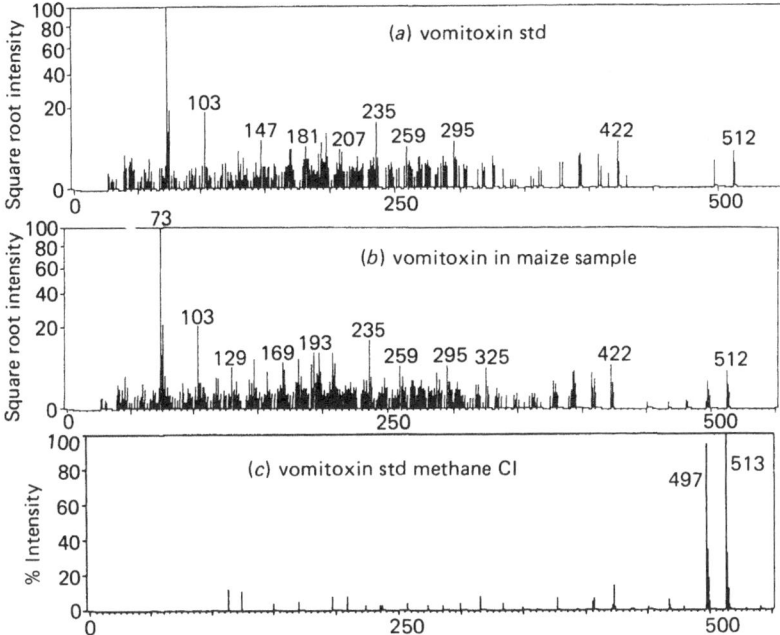

Fig. 11.4. Comparison of (*a*) reference mass spectrum of TMS-deoxynivalenol and (*b*) peak identified in flaked maize extract. Spectra obtained at 70 eV are displayed as square root intensities for ease of comparison of high mass ions. (*c*) shows chemical ionisation spectrum for TMS-deoxynivalenol with methane reagent gas.

pseudo-molecular ion at m/z = 513 and little other fragmentation. This enables much easier assignment of the molecular weight and in conjunction with the electron impact spectrum a greater degree of certainty of positive identification.

As an alternative to obtaining a full spectrum of a trichothecene, the mass spectrometer can also be operated as a detector where one or more ions thought to be specific to the compound of interest are monitored. This mode of operation is ideal for quantitative work, and the sensitivity which can be attained is considerably better than that from operation in a full scanning mode. Also the sensitivity is better than can be achieved with most other GC dectectors, added to which the specificity is considerably higher. An example of single ion monitoring of TMS deoxynivalenol is illustrated in Fig. 11.5 showing for m/z = 512 a single peak in the chromatogram at the deoxynivalenol retention time.

Although this static mode of operation affords maximum sensitivity at optimum resolution, it does limit the GC assay to one specified

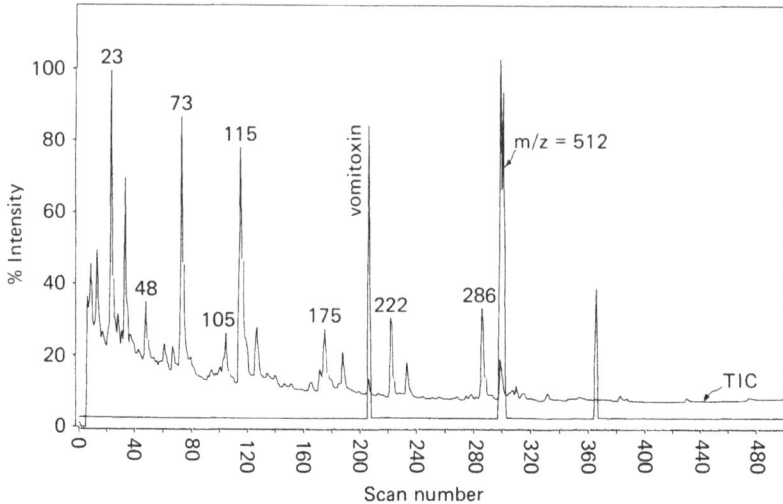

Fig. 11.5. Total ion chromatogram (TIC) and selected ion monitoring (SIM) for m/z = 512 for flaked maize extract. GC conditions as for Fig. 11.3.

trichothecene (unless a common fragment ion can be chosen) and excludes the use of an internal standard. An alternative which overcomes these shortcomings but has a slightly reduced sensitivity is multiple ion detection, in which the mass spectrometer operates in a switched mode for a number of pre-selected masses. This operation is best carried out under computer control, and for a multi-trichothecene screen one might either monitor for example eight ions simultaneously for a complete GC run or monitor chosen pairs of different ions characteristic of each trichothecene while switching between pairs for different retention time windows during the course of a run. The latter mode of operation is illustrated in Fig. 11.6 for a mixture of six TMS trichothecene reference standards, in each case the molecular ion plus one other characteristic fragment ion being monitored for a three-minute retention time window. In this instance because of the closeness of retention times, two batches of six ions are in fact monitored during the course of the one run.

Using selected ion monitoring a sensitivity of the order of 10–20 pg for components on-column should be easily attainable for standards, which, by calculation, equates with a contamination of 0.1 to 0.2 μg/kg in the commodity. In practice, however, the sensitivity of the mass spectrometer at these levels becomes limited by 'chemical noise' even in

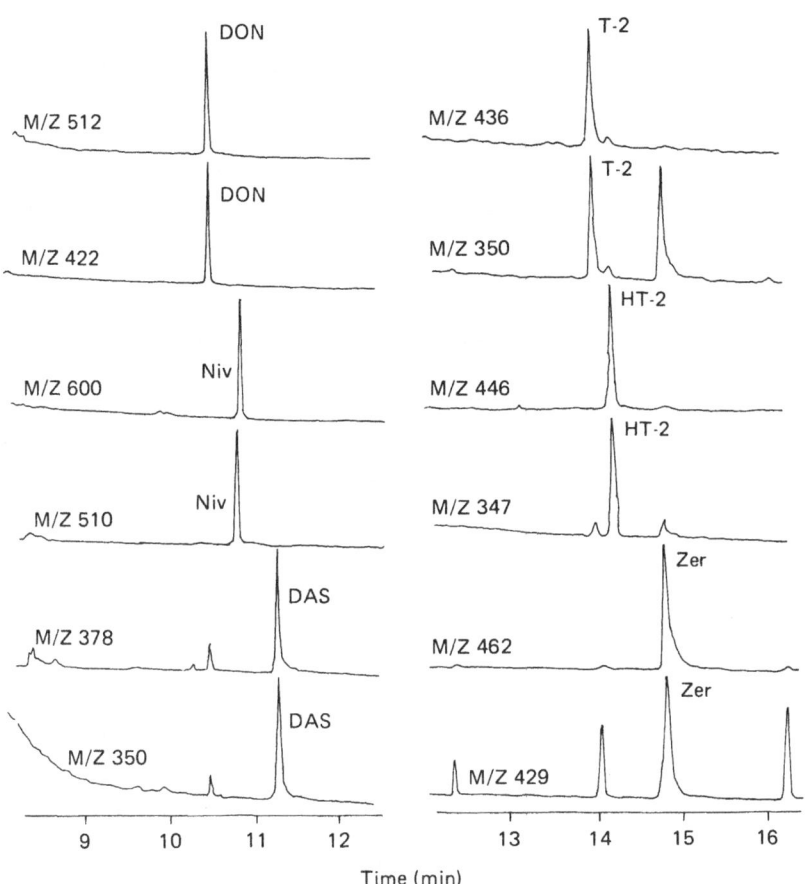

Fig. 11.6. MID chromatogram for TMS trichothecene mixture of reference standards. DON, deoxynivalenol; Niv, nivalenol, DAS, diacetoxyscirpinol. T-2, T-2 toxin, HT-2, HT-2 toxin, Zer, zearalenone. Mass spectrometer under computer control monitoring six ions as indicated for retention time window 9 to 12 min, and a different set of six ions for retention time window 12 to 16 min.

a cleaned-up sample and therefore a working limit of detection would probably be around 10 to 20 μg/kg.

Internal standards

Although there is little doubt that the specificity of mass spectrometry makes for very reliable qualitative assays, the instrument reproducibility is generally not as good as with other GC detectors, and additional difficulties specific to trichothecenes have been experienced in quantitative work both in the author's own laboratory and elsewhere

(Pareles, Collins & Rosen, 1976). Although successive injections of calibration standards appear to give essentially reproducible results, spiked extracts have led to stepwise changes in instrument response during the course of a day's operation.

The normal way of overcoming these difficulties is to incorporate an internal standard and the preferred choice would be a deuterated analogue of the compound of interest. The advantage of this type of internal standard is that it is structurally identical and co-chromatographs with the compound of interest, but using multiple ion detection can be assayed separately because of its mass difference. For the trichothecenes the synthesis of a deuterated internal standard is not a feasible chemical proposition, and the only realistic alternative is to prepare the deuterated TMS derivative (Pareles *et al.*, 1976; Collins & Rosen, 1979). This has the disadvantage that there is a risk of 'scrambling' of the deutero atoms and the additional shortcoming that the deuterated TMS reagent is too expensive for routine usage.

Other possibilities are to use a closely chromatographing non-mycotoxin organic compound as an internal standard, e.g. methoxychlor (Romer, Boling & MacDonald, 1978) or to synthesise a non-naturally occurring trichothecene, e.g. *iso*-T-2 toxin (Stahr *et al.*, 1981). The latter appears to be the best compromise and in the author's own laboratory mixtures of both scirpenetriol and deuterated diacetoxyscirpenol have been prepared from diacetoxyscirpenol and are incorporated as internal standards. The advantage of adding an internal standard is that not only is the quantitative precision of the mass spectrometry greatly improved but if the internal standard is incorporated at the extraction stage of the assay an internal reference is provided as a monitor throughout the recovery at the clean-up and derivatisation stages.

Analysis of zearalenone

Although the sophisticated GC–MS techniques described for the trichothecenes can be applied equally successfully to the analysis of TMS derivatised zearalenone, the native fluorescent properties of this compound make it particularly amenable to detection methods available for t.l.c. and h.p.l.c. analysis. The analysis of this oestrogenic *Fusarium* mycotoxin will therefore be outlined primarily in relation to these techniques.

Extraction and clean-up

Various polar solvents ranging from aqueous acetonitrile (Patterson & Roberts, 1979; Hunt, Bourdon & Crosby, 1978), aqueous methanol (Thomas, Eppley & Trucksess, 1975), mixtures of chloroform/ethanol (Cohen & Lapointe, 1980), water/ethyl acetate (Mirocha, Schaverhamer & Pathre, 1974) and chloroform/phosphoric acid (Josefsson & Moller, 1979) have been described as zearalenone extractants using normal methods of homogenisation in a blender or continuous shaking of the finely ground commodity with the solvent. Liquid/liquid partition to remove lipids (Patterson & Roberts, 1979; Hunt *et al.*, 1978), protein precipitation, silica-gel or Sep-Pak column chromatography, dialysis, size exclusion chromatography, or a combination of these techniques (Holder, Nony & Bowman, 1977) normally follow as clean-up stages. A particularly useful step common to a number of methods has utilised the phenolic nature of zearalenone and incorporated alkaline extraction as a purification stage (Mirocha *et al.*, 1974; Ware & Thorpe, 1978).

As with most extraction and clean-up methods described in the literature for mycotoxins selection of one method over another is largely a matter of personal preference. Hence, if the size of the commodity sample initially extracted is chosen to be arbitrarily large and one is prepared to carry out effective (but often lengthy) clean-up, this can be traded off against ultimate poor detector sensitivity and specificity; in other words, an increase in sample size combined with correspondingly more clean-up will offset a poorer detector sensitivity without a sacrifice in the overall ability of the method to detect the required μg/kg level of contamination. Hence the clean-up chosen will be dependent on the separation and detection method to be used, the overall sensitivity needed, and the type of commodity to be analysed, as some materials invariably contain larger amounts of interfering substances than others.

Separation and quantification by TLC and HPLC

On silica gel t.l.c. using chloroform methanol (93.7) zearalenone appears under UV light as a blue/green fluorescent spot with an R_f of 0.6. Approximately 0.1 μg of zearalenone is visually detectable on the plate which, for most clean-up procedures, would correspond to a limit of detection of 50 μg/kg in the contaminated commodity. Additional confirmation of the identity of zearalenone can be obtained by the use of various spray reagents, for example, with aluminium chloride it continues to be observed as a blue spot with enhanced fluorescence (Eppley *et al.*, 1974) and with bis-diazotized benzidine (Malaiyandi, Barrette & Wavrock, 1976) as a bright red spot with a claimed sensitivity of 1.5 ng per spot. As with most extracts from biological materials a large number of other non-mycotoxin components are usually evident by t.l.c. and hence for most commodities the use of 2-dimensional t.l.c. would be strongly advocated, to improve the separation of zearalenone from other potential interferences.

For h.p.l.c. analysis both normal and reverse phase techniques have been employed and a useful critical comparison of the various chromatographic conditions has been published by Scott (1981). In general, combinations of methanol/water are employed isocratically in reverse phase systems, whilst chloroform-based mobile phases incorporating either iso-octane, methanol, acetonitrile or acetic acid have been employed in normal phase systems. Although detection can be adequately carried out by UV at 254 nm, fluorescence is the detection method of choice, being preferable both in terms of sensitivity and specificity. A typical chromatogram illustrating the h.p.l.c. analysis of zearalenone in contaminated maize is shown in Fig. 11.7, indicating the apparent cleanliness of the extract in this mode of detection. In general terms sensitivities from low nanogram down to sub-nanogram (corresponding to 2 to 50 μg/kg) should be easily achievable by fluorescence h.p.l.c. and even better limits can be achieved using laser fluorimetry (Siebold, Karny & Zare, 1979), although this is as yet not an instrumental technique which is commercially available.

One of the advantages of h.p.l.c. determinations compared with t.l.c. is the ease of quantification and the improved precision of peak area measurement as opposed to subjective estimations of the intensity of the colour of fluorescence of t.l.c. spots. Also an added dimension of confirmation by h.p.l.c. in the fluorescence mode can be obtained by determining ratios of excitation wavelengths, e.g. 236:255, 236:274 and 236:314 (Cohen & Lapointe, 1980; Ware & Thorpe, 1978). The peak height ratios of suspected zearalenone in the samples should agree to within about 5% of that obtained for authentic zearalenone. With modern microprocessor controlled fluorescence h.p.l.c. systems it is comparatively easy to achieve this facility of automatically switching wavelengths.

Analysis of other *Fusarium* metabolites

In addition to the trichothecenes and zearalenone, two other compounds, i.e. butenolide and moniliformin (see Fig. 11.1), have been

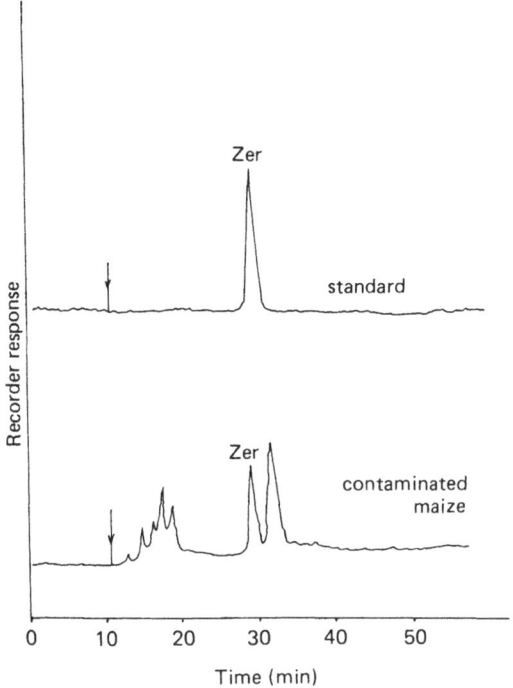

Fig. 11.7. Chromatogram illustrating zearalenone standard and zearalenone in naturally contaminated maize extract. HPLC conditions: column, Spherisorb ODS (5 μm) 25 cm × 4.9 mm ID; mobile phase, methanol–water (60 + 40); flow rate 1.0 ml/min; fluorescence detector excitation 312 nm and emission measured at 445 nm.

identified as *Fusarium* metabolites. Both have been extracted from cereals by an identical procedure to that employed for the trichothecenes (Kamimura *et al.*, 1981) although fractionated from the trichothecenes during the Amberlite XAD chromatographic clean-up stage. By t.l.c. both compounds can be visualised using, 2,4-dinitrophenylhydrazine spray and heating at 110 °: moniliformin then appears as a red-brown spot and butenolide as a yellow spot. Some added degree of confirmation can be obtained by micro-scale preparation of the 2,4-dinitrophenylhydrazone derivatives prior to t.l.c. and, using scanning densitometry, limits of detection of 30 and 50 μg/kg have been claimed for butenolide and moniliformin respectively (Kamimura *et al.*, 1981).

For moniliformin in maize samples, a simple extraction procedure with water has been employed and, after centrifugation and filtration or a Sephadex A50 clean-up, a direct h.p.l.c. analysis was carried out. H.p.l.c. was by either ion-exchange with sodium dihydrogen phosphate solution or by ion-pair chromatography (using sodium phosphate buffer in methanol), in both cases detection being by low wavelength UV. Coincidence of retention times with the standard in the two h.p.l.c. systems was taken as being confirmation of identity of moniliformin (Thiel, Meyer & Marasas, 1982), and the limit of detection was reported as around 1 mg/kg.

References

Bennett, G. A., Peterson, R. E., Plattner, R. D. & Shotwell, O. L. (1981). Isolation and purification of deoxynivalenol and a new trichothecene by high pressure liquid chromatography. *Journal of the American Oil Chemists Society*, **58**, 1002A-5A.

Cohen, H. & Lapointe, M. R. (1980). Sephadex LH-20. Clean-up, high pressure liquid chromatographic assay and fluorescence detection of zearalenone in animal feeds. *Journal of the Association of Official Analytical Chemists*, **63**, 642–6.

Collins, G. J. & Rosen, J. D. (1979). Gas-liquid chromatographic/mass spectrometric screening method for T-2 toxin in milk. *Journal of the Association of Official Analytical Chemists*, **62**, 1274–80.

Diebold, G. J., Karny, N. & Zare, R. N. (1979) Determination of zearalenone in corn by laser fluorimetry. *Analytical Chemistry*, **51**, 67–9.

Eppley, R. M., Stoloff, L., Trucksess, M. W. & Chung, C. W. (1974). Survey of corn for *Fusarium* toxins. *Journal of the Association of Official Analytical Chemists*, **57**, 632–5.

Forsyth, D., Yoshizawa, T., Morooka, N. & Tuite, J. (1977). Emetic and refusal activity of deoxynivalenol to swine. *Applied Environmental Microbiology*, **34**, 547–51.

Holder, C. L., Nony, C. R. & Bowman, M. C. (1977). Trace analysis of zearalenone and/or zearalenol in animal chow by high pressure liquid chromatography and gas–liquid chromatography. *Journal of the Association of Official Analytical Chemists*, **60**, 272–8.

Hunt, D. C., Bourdon, A. T. & Crosby, N. T. (1978). Use of high performance liquid chromatography for the identification and estimation of zearalenone, patulin and penicillic acid in food. *Journal of the Science of Food and Agriculture*, **29**, 239–44.

Ilus, T., Niku-Paavola, M. L. & Enari, T-M. (1981). Chromatographic analysis of *Fusarium* toxins in grain samples. *European Journal of Applied Microbiology and Biotechnology*, **11**, 244–247.

Josefsson, E. & Moller, T. (1979). High pressure liquid chromatographic determination of ochratoxin A and zearalenone in cereals. *Journal of the Association of Official Analytical Chemists*, **62**, 1165–8.

Kamimura, H., Nishijima, M., Yasuda, K., Saito, K., Ibe, A., Nagayama, T., Ushiyama, H. & Naoi, Y. (1981). Simultaneous detection of several *Fusarium* mycotoxins in cereals, grains and foodstuffs. *Journal of the Association of Official Analytical Chemists*, **64**, 1067–73.

Malaiyandi, M., Barrette, J. P. & Wavrock, P. L. (1976). Bis-diazotized benzidine as a spray reagent for detecting zearalenone on thin layer chromatoplates. *Journal of the Association of Official Analytical Chemists*, **59**, 959–62.

Mirocha, C. J., Schaverhamer, B. & Pathre, S. V. (1974). Isolation, detection and quantitation of zearalenone in maize and barley. *Journal of the Association of Official Analytical Chemists*, **57**, 1104–10.

Pareles, S. R., Collins, G. J. & Rosen, J. D. (1976). Analysis of T-2 toxin (and HT-2 toxin) by mass fragmentography. *Journal of Agricultural and Food Chemistry*, **24**, 872–5.

Pathre, S. V. & Mirocha, C. J. (1977). Assay methods for trichothecenes and review of their natural occurrence. In *Mycotoxins in Human and Animal Health*, ed. J. V. Rodericks, C. W. Hesseltine and M. A. Mehlman, pp. 229–53. Illinois: Pathotox Publishers, Inc.

Patterson, D. S. P. & Roberts, B. A. (1979). Mycotoxins in animal feedstuffs: sensitive thin layer chromatographic detection of aflatoxin, ochratoxin A, sterigmatocystin, zearalenone and T-2 toxin. *Journal of the Association of Official Analytical Chemists*, **62**, 1265–7.

Romer, T. R., Boling, T. M. & MacDonald, J. L. (1978). Gas-liquid chromatographic determination of T-2 toxin and diacetoxyscirpenol in corn and mixed feeds. *Journal of the Association of Official Analytical Chemists*, **61**, 801–8.

Sano, A., Asabe, Y., Takitani, S. & Ueno, Y. (1982). Fluorodensitometric determination of trichothecene mycotoxins with nicotinamide and 2-acetylpyridine on a silica gel layer. *Journal of Chromatography*, **235**, 257–65.

Scott, P. M. (1981). Liquid chromatography in the analysis of mycotoxins. In *Trace Analysis, Volume 1*, ed. J. F. Lawrence, pp. 193–266. London: Academic Press.

Scott, P. M., Lau, P-Y. & Kanhere, S. R. (1981). Gas chromatography with electron capture and mass spectrometric detection of deoxynivalenol in wheat and other grains. *Journal of the Association of Official Analytical Chemists*, **64**, 1364–71.

Scott, P. M., Lawrence, J. W. & van Walbeek, W. (1970). Detection of mycotoxins by thin-layer chromatography: application to screening of fungal extracts. *Applied Microbiology*, **20**, 839–42.

Stahr, H. M., Hyde, W., Lerdal, D. & Pfeiffer, R. (1981). Trichothecene mycotoxin analysis for veterinary diagnostic toxicology. *Abstracts of the 95th Annual Meeting of the AOAC*, No. 191, p. 65.

Takitani, S., Asabe, Y., Kato, T., Suzuki, M. & Ueno, Y. (1979). Spectrodensitometric determination of trichothecene mycotoxins with 4-(p-nitrobenzyl)pyridine on silica gel thin-layer chromatograms. *Journal of Chromatography*, **172**, 335–42.

Thiel, P. G., Meyer, C. J. & Marasas, W. F. O. (1982). Natural occurrence of moniliformin together with deoxynivalenol and zearalenone in Transkeian corn. *Journal of Agricultural and Food Chemistry*, **30**, 308–12.

Thomas, F., Eppley, R. M. & Trucksess, M. W. (1975). Rapid screening method for aflatoxins and zearalenone in corn. *Journal of the Association of Official Analytical Chemists*, **58**, 114–16.

Ueno, Y., Saito, N., Ishii, K., Sakai, K., Tounoda, H. & Enomoto, M. (1973). Biological and chemical detection of trichothecene mycotoxins of *Fusarium* spp. *Applied Microbiology*, **25**, 699–704.

Vesonder, R., Ciegler, A., Jensen, A., Rohwedder, W. & Weisleder, D. (1976). Co-identity of the refusal and emetic principle from *Fusarium* infected corn. *Applied Environmental Microbiology*, **31**, 280–5.

Ware, G. W. & Thorpe, C. W. (1978). Determination of zearalenone in corn by high pressure liquid chromatography and fluorescence detection. *Journal of the Association of Official Analytical Chemists*, **61**, 1058–62.

12
The biosynthesis of *Fusarium* mycotoxins

M.O.MOSS

Department of Microbiology, University of Surrey, Guildford, Surrey GU2 5XH, U.K.

Introduction

The fungi associated with terrestrial habitats have evolved two quite distinct types of morphology for the period of vegetative growth, giving rise to the two groups conveniently referred to as yeasts and moulds. It is striking that it is the latter which are so prominently associated with the production of secondary metabolites. Diversity of secondary metabolite production is not associated with yeasts, indeed Harrison (1970) makes no mention of any such compound during a review of miscellaneous products from yeasts.

We see a somewhat analagous situation amongst the heterotrophic procaryotes where it is the filamentous actinomycetes that are especially associated with the production of a remarkable array of secondary metabolites.

The development of mycelium represents a simple but efficient means of utilising complex organic matter in the solid phase, giving intimate contact with the substrate, and allowing the localised secretion of hydrolytic enzymes and absorption of the small molecules released. Of course water needs to be available for all these processes, and water activity is one of the major factors controlling mould growth and metabolism, but the moulds can grow and spread without the need for an aquatic environment.

The mycelium of one mould is much like that of another and the marvellous diversity of speciation is apparent from the diversity of propagules, formed both sexually and asexually, for dispersal and survival. Secondary metabolite formation is another aspect of species diversity although mycologists are justifiably wary about incorporating such information into their formal taxonomy.

Although it is not yet possible to rationalise the observation that it is the filamentous fungi and not the single celled fungi which produce the majority of fungal secondary metabolites, the following observations may be relevant.

The hypha is a polarised structure with a well-defined growing tip (Burnett, 1979) leaving behind cytoplasm that may become increasingly vacuolated (Robinson, Park & McClure, 1969). In single celled organisms, including the yeasts, the individual cell may die in isolation from its fellows except through the agency of an aqueous environment. In the filamentous fungi there is still the possibility of transport of materials from aging cytoplasm to the growing tip.

Through careful studies such as those of Borrow *et al.* (1961) on *Gibberella fujikuroi* (Sawada) Ito, it has been shown that, in the absence of any nutritional requirements, filamentous fungi may show a balanced phase of growth during which the gross composition of the mycelium remains constant.

In the case of *G. fujikuroi* it was shown that, following the exhaustion of nitrogen and in the presence of sufficient glucose, the gross composition of the mycelium changes. For example the lipid content increased from 6% to 45% of the dry weight of mycelium in one particular group of experiments.

Mycotoxins, including those produced by *Fusarium*, may clearly be included in the group of fungal metabolites conveniently referred to as secondary metabolites, the production of which is associated with the later stages of growth of batch cultures often called the stationary or maintenance phase.

Secondary metabolism

If primary metabolism is considered to be the activity and interaction of catabolic, anabolic, and anaplerotic pathways leading to the formation and differentiation of biomass then the area of metabolism with which we are concerned is secondary in the sense that:

(i) It usually occurs after a phase of balanced primary metabolism.

(ii) Whereas primary metabolism involves pathways and intermediates common to many forms of life, the production of particular secondary metabolites is frequently species, or even strain, specific and is often very sensitive to the physio-chemical parameters of the environment.

(iii) It has not been possible to rationalise roles of individual secondary metabolites in the biology of the producing organisms, although it may be useful to think about a role for secondary metabolism as a process (Bu'Lock, 1980).

(iv) The active intermediates accepted as the precursors for the biosynthesis of secondary metabolites are themselves products or intermediates of primary metabolism.

Because many secondary metabolites have biological activity as antibiotics, phytotoxins and mycotoxins, it is tempting to associate this activity with a role in the biology and ecology of the producing organisms. This may be possible in individual cases but it is certainly not possible to generalise. The production of secondary metabolites often occurs at the time when a mould is undergoing morphogenesis leading to the production of spores. Again it is tempting to look for a role for the metabolites in the process of morphogenesis. It has been suggested that zearalenone may be involved in the regulation of sexual reproduction in *Gibberella zeae* (Schw.) Petch (Wolf & Mirocha, 1973) although the possibility that zearalenone should be considered as a fungal hormone has been viewed with caution (Bu'Lock, 1976). Nevertheless it has been demonstrated that both perithecial development and zearalenone biosynthesis are stimulated by cyclic AMP (Wolf & Mirocha, 1977) and that *Gibberella zeae* produces a protein with a binding site for zearalenone (Inaba & Mirocha, 1979) which the authors suggest may play a role in morphogenesis. Any hypothesis concerning the possibility that zearalenone itself, rather than the process by which it is produced, may have a role in the biology of the producing organisms will need to account for the observation that *Fusarium culmorum* (W. G. Smith) Sacc. and *F. sporotrichioides* Sherb. (neither of which are associated with a perfect stage) are reported to produce zearalenone.

In general, the secondary metabolism of the filamentous fungi is complex and many strains are able to produce numerous metabolites, often involving quite different pathways, under apparently the same growth conditions. They may, in fact, be produced sequentially, either because they represent different stages in a single biosynthetic pathway, or because the production of each is controlled by the physiological status of the mould, which itself is time dependent in a batch culture. One of the best-studied examples of the latter is the study of bikaverin and gibberellin production in *Gibberella fujikuroi* by Bu'Lock *et al.* (1974). They showed that bikaverin, which is a polyketide metabolite requiring both malonyl and acetyl coenzyme A, is produced at an earlier stage than gibberellin, a diterpene whose biosynthesis requires

Table 12.1. *Trichothecene derivatives produced by moulds other than* Fusarium

Parent trichothecene	Oxidation pattern	Genera
Trichodermol	4-OH	*Myrothecium* *Trichoderma*
Trichothecolone	4-OH; 8-oxo	*Trichothecium*
Verrucarol	4, 15-diOH	*Myrothecium* *Stachybotrys* *Verticimonosporium* *Cylindrocarpon*
Crotocin	4-OH; 7,8 epoxy	*Cephalosporium*

Table 12.2. *Distribution of some metabolites amongst species of* Fusarium

Species	Parent trichothecene					Moniliformin	Zearalenone	Gibberellins	Butenolide
	3,4,7,15-OH; 8-oxo	3,7,15-OH; 8-oxo	3,4,8,15-OH	3,4,15-OH	3,15-OH				
F. graminearum	+	+	−	−	−	−	+	−	−*
F. equiseti	+	−	+	+	−	+	−	−	−
F. semitectum	+	−	+	+	−	+	−	−	−
F. nivale	+	−	−	+	+	−	−	−	+
F. culmorum	−	+	−	−	+	−	+	−	−
F. acuminatum	−	−	+	+	−	+	−	−	−
F. sporotrichioides	−	−	+	+	−	−	+	−	−
F. lateritium	−	+	−	+	−	−	−	−	−
F. tricinctum	−	−	+	+	−	−	−	−	−
F. solani	−	−	+	−	−	−	−	−	−
F. poae	−	−	+	−	−	−	−	−	−
F. heterosporum	−	−	+	−	−	−	−	−	−
F. sulphureum	−	−	−	+	−	−	−	−	−
F. sambucinum	−	−	−	+	−	−	−	−	−
F. oxysporum	−	−	−	−	−	+	−	−	−
F. moniliforme	−	−	−	−	−	+	−	+	−

Note: *produces 4-acetamido 2 butenoic acid.

mevalonate which itself is produced from acetyl coenzyme A. They also demonstrated that, by controlling the specific growth rate of the mould using a chemostat, the two metabolites are produced optimally at two different growth rates, bikaverin at $0.05\,h^{-1}$ and gibberellin at $0.01\,h^{-1}$ (see Fig. 13.4).

Fusarium mycotoxins

The majority of toxic metabolites produced by species of *Fusarium* belong to the trichothecene group of sesquiterpenes. Although those trichothecenes produced by species of *Fusarium* seem to be confined to this genus, there are a number of other trichothecene metabolites produced by fungi in other genera (Table 12.1).

The production of the quite unrelated mycotoxin zearalenone, on the other hand, seems to be confined not only to members of the genus *Fusarium*, but to a limited number of closely related species (Table 12.2).

There are two more well-documented toxic metabolites, a butenolide produced by strains of *Fusarium nivale* (Fr.) Cesati isolated from the grass *Festuca arundinacea* Schreb. (Yates *et al.*, 1968), and moniliformin, originally isolated from a strain of *F. moniliforme* Sheldon (Cole *et al.*, 1973) but now reported from a number of other species of *Fusarium*. Although these two metabolites have very simple structures (Fig. 12.1) nothing has yet been reported concerning their biosynthesis. Moniliformin has been synthesised in the laboratory (Springer *et al.*, 1974) but the route used, involving the addition of dichloroketene to ethoxyacetylene (Fig. 12.2), does not throw any light on how it may be produced by a mould.

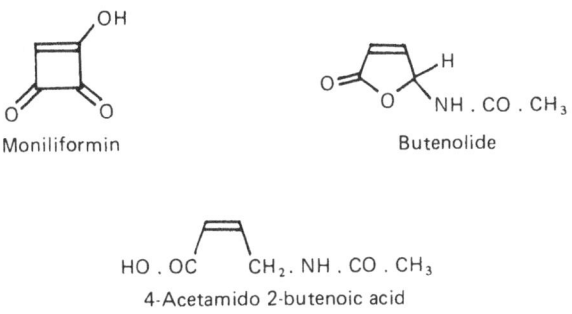

Fig. 12.1. Structures of three simple metabolites of species of *Fusarium*.

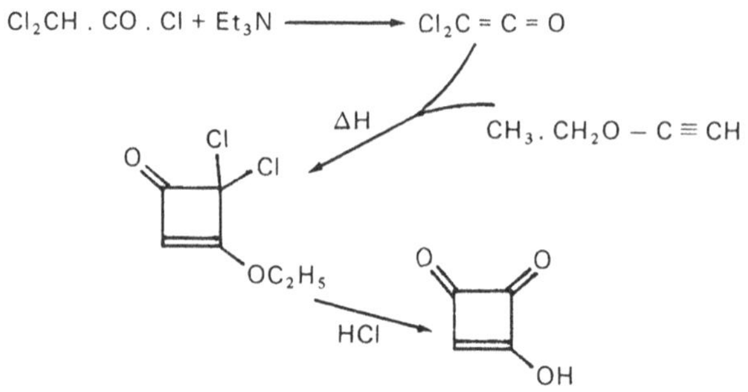

Fig. 12.2 The chemical synthesis of moniliformin.

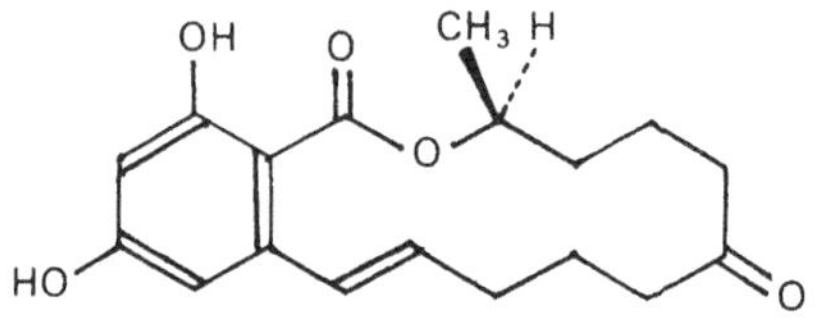

Zearalenone

Fig. 12.3. The structure of zearalenone, an oestrogenic mycotoxin produced by some species of *Fusarium*.

The compound 4-acetamido 2-butenoic acid has been isolated from a strain of *Fusarium graminearum* (Vesonder *et al.*, 1977). It is not toxic but the authors suggest that it may be an intermediate in the biosynthesis of butenolide. It is certainly possible that the two compounds have related biosynthetic pathways.

The biosynthesis of zearalenone

An examination of the structure of zearalenone shows it to be an almost perfect candidate as a polyketide derived metabolite with oxygen functions in the positions expected to be derived from the carboxyl group of acetate (Fig. 12.3).

The biosynthesis of polyketide derived metabolites proceeds by the reaction of the active methylene group in malonyl coenzyme A with the carboxyl group of an acyl coenzyme A which, in the first step, is the acetyl derivative. Thus zearalenone would require one molecule of acetyl CoA and eight of malonyl CoA (Fig. 12.4). These reactions are entirely analogous with those involved in the biosynthesis of fatty acids

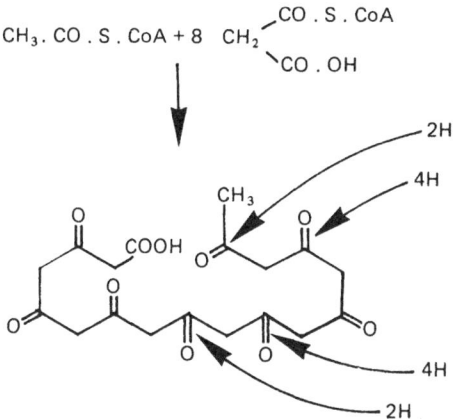

Fig. 12.4. The polyketide route to zearalenone.

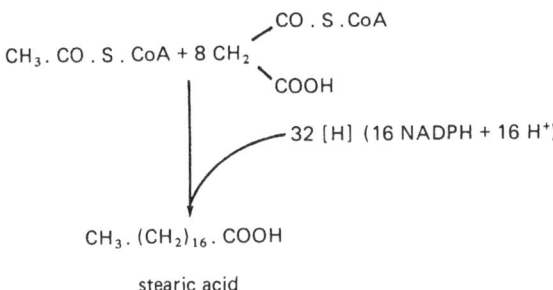

stearic acid

Fig. 12.5. The biosynthesis of C_{18} fatty acids.

with the difference that the latter requires a number of NADPH dependent reductions (Fig. 12.5).

In the fungi fatty acid biosynthesis is mediated by a single multi-enzyme complex (Lynen, 1967). It is likely that polyketide biosynthesis also involves a similar multi-enzyme complex and, indeed, a 6-methyl salicyclic acid synthetase has been isolated and purified from *Penicillium patulum* Bain. (Dimroth, Walter & Lynen, 1970).

Assuming that the polyketide synthetases are distinct from fatty acid synthetase, and not simply degenerate forms of it which have lost their reducing capacity, it is possible that they may have evolved through gene duplication of those genes whose products form the fatty acid synthetase, with subsequent mutation of the second copy.

Such a synthetase has not been isolated for zearalenone although it has been shown that both ^{14}C-labelled acetate and malonate are

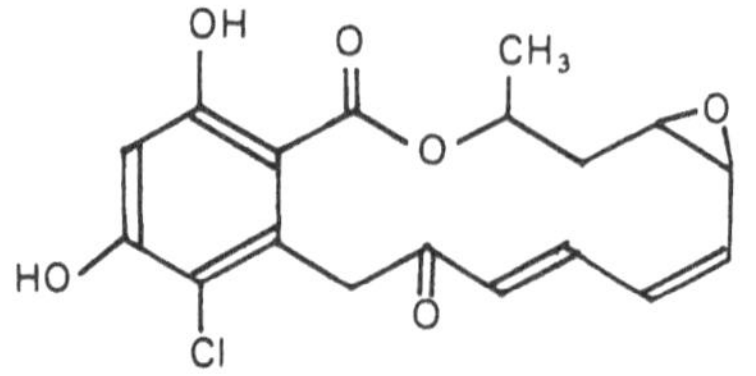

Fig. 12.6. The structure of monorden (=radicicol), a metabolite of *Monosporium bonorden* and *Nectria radicicola*.

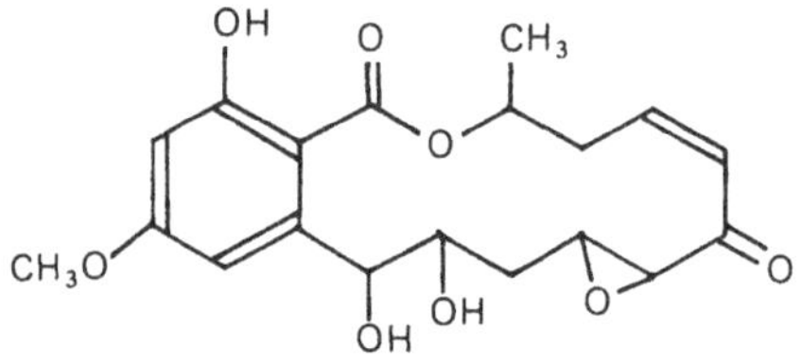

Fig. 12.7. The structure of hypothemycin, a metabolite of *Hypomyces trichothecoides*.

incorporated into zearalenone to give rise to the expected labelling pattern, and that malonate inhibits the incorporation of acetate (Steele, Lieberman & Mirocha, 1974).

Although yields as high as 38 g per kilogram of substrate have been obtained on solid media containing maize and rice it seems that defined liquid media fail to support the production of significant quantities of zearalenone (Mirocha, Christensen & Nelson, 1971). Using a solid medium of moistened, polished rice, Hagler & Mirocha (1980) have been able to obtain efficient incorporation of $1^{14}C$ acetate to give high yields of labelled zearalenone required for experimental studies.

Zearalenone is the most reduced member of the small group of undecyl resorcylic acid derivatives known to be produced by fungi. The antifungal metabolite known both as monorden and radicicol (Fig. 12.6) was first isolated from a soil organism referred to as *Monosporium* Bonorden (Delmotte & Delmotte-Plaquee, 1953) and characterised by McCapra *et al.* (1964). It was subsequently isolated and characterised from *Nectria radicicola* Gerlach & Nilsson (*Cylindrocarpon radicicola* Wr.) by Mirrington *et al.* (1964). This compound is described as having a low toxicity but potent sedative properties. A related metabolite, isolated from *Hypomyces trichothecioides* Tubaki and called (+) hypothemycin (Fig. 12.7), has been described by Nair & Carey (1980).

Table 12.3. *Production of lipids by* Fusarium

Species	Lipid (% w/w)
F. bulbigenum*	50
F. graminearum	31
F. lini*	35
F. lycopersici*	35–40
F. oxysporum	29–34

Source: From Ratledge, 1982.
Note: * These names may be synonyms of *F. oxysporum.*

An examination of the monoketide chain which forms the skeleton of zearalenone shows it to have two distinct regions. Most of the leading pentaketide section is reduced in a beautifully symmetrical manner about the ketone function requiring the equivalent of six molecules of NADPH (Fig. 12.4). The final tetraketide section retains its original oxidation level undergoing aldol condensation to form the resorcylic acid component of the zearalenone molecule. It would be very interesting to know whether there is any link between the biosynthetic machinery leading to the formation of the C_{18} fatty acids and zearalenone synthetase. Certainly, a number of species of *Fusarium*, including *F. graminearum*, are known to accumulate quite large quantities of lipids (Table 12.3). In a study of the lipids from 15 species of *Fusarium*, Gorbik, Pidoplichko & Loiko (1980) showed that some species produced as much as 35–50% of their dry weight as lipid and that, of this lipid, 70–80% was in the form of triglycerides, the major fatty acids in which were the C_{18} acids stearic, oleic, linoleic and linolenic acids.

Biosynthesis of trichothecenes

Probably the first material now known to be a trichothecene was glutinosin isolated by Brian & McGowan (1946) from a fungus which they referred to as *Metarhizium glutinosum* S. Pope, although it was probably *Myrothecium verrucaria* (Alb. and Schw.: Fr.) Ditmar: Fr. This material was subsequently shown to be a mixture of verrucarins A and B (Grove, 1968). Trichothecin, the isocrotonic ester of trichothecolone, was isolated soon after as an antifungal metabolite of *Trichothecium roseum* Link (Freeman & Morrison, 1948).

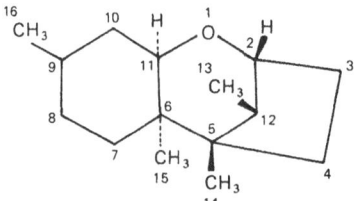

Fig. 12.8. Trichothecan, the parent structure of all the trichothecenes, and the numbering system adopted for identifying positions around the tricyclic skeleton.

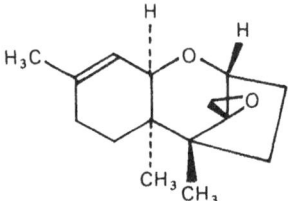

Fig. 12.9. 12, 13-epoxytrichothec 9-ene, the nucleus common to the majority of trichothecene metabolites.

Thus, although their structures had not been correctly defined, trichothecenes were an established group of biologically active mould metabolites before their association with the genus *Fusarium*. It was really an extension of the interest in gibberellic acid production by *Fusarium moniliforme* Sheld. that led workers of the ICI Akers Research Laboratories to the discovery of the phytotoxic metabolite diacetoxyscirpenol from strains of *F. equiseti* (Corda) Sacc., including isolates referred to as *F. scirpi* Lamb. & Fautr. which are now accepted as synonymous with *F. equiseti* (Brian *et al.*, 1960).

The exact nature of the ring system in this now extensive family of compounds was established by X-ray crystallography of the *p*-bromobenzoate of trichodermol, the parent alcohol of trichodermin isolated from *Trichoderma viride* Pers.: Fr. (Abrahamsson & Nilsson, 1964). It was agreed that they should all be considered as derivatives of a hypothetical parent compound trichothecan (Fig. 12.8) and, indeed, all but two are derivatives of 12, 13 epoxy trichothec 9-ene (Fig. 12.9).

It was evident from the earliest stages to those people working on the structure of the trichothecenes that they are sesquiterpenes and Jones &

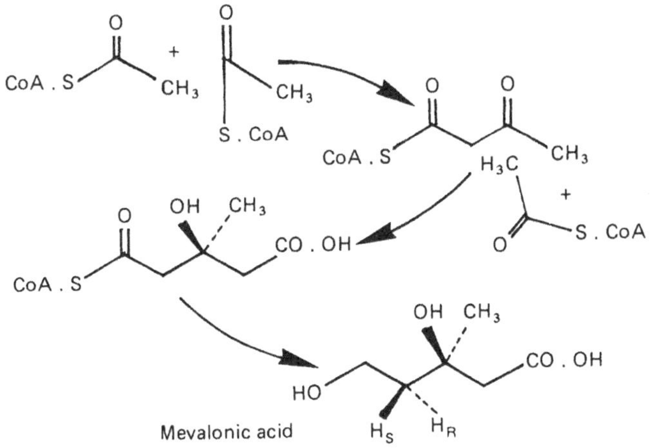

Fig. 12.10. The biosynthesis of mevalonic acid.

Lowe (1960) demonstrated that labelled mevalonate is incorporated into the trichothecene nucleus (and not into the isocrotonyl side chain) of trichothecin. Referring to the sesquiterpenes, Arigoni (1975) comments that 'Nature has been particularly prodigal in this area in playing variations on a known theme, and it seems that almost every conceivable structural possibility, including some which would have defied the ingenuity of even the most daring organic chemist, has been put into practice by living organisms.' The trichothecenes represent a small part of this diversity of structure but their importance has ensured that they have been studied in great detail. Much of this detailed work on the biosynthesis of the trichothecenes has been reviewed by Tamm & Breitenstein (1980) and it is appropriate here just to make some general observations.

Mevalonate itself is produced from three molecules of acetyl coenzyme A (Fig. 12.10). The possibility that malonyl coenzyme A may be involved has been suggested but is considered to be unlikely (Charlwood & Banthorpe, 1978).

Acetyl coenzyme A is thus a key intermediate in both the production of polyketide and mevalonate derived secondary metabolites. It is worth noting that acetyl coenzyme A is not only a key intermediate, entering the tricarboxylic acid cycle (TCA) for the generation of energy, but also, with the help of the glyoxylate pathway and other anaplerotic reactions, it is an intermediate in the production of such TCA generated metabolites as glutamate and aspartate. Acetyl coenzyme A may also be subsequently involved in the biosynthesis of a further range of amino

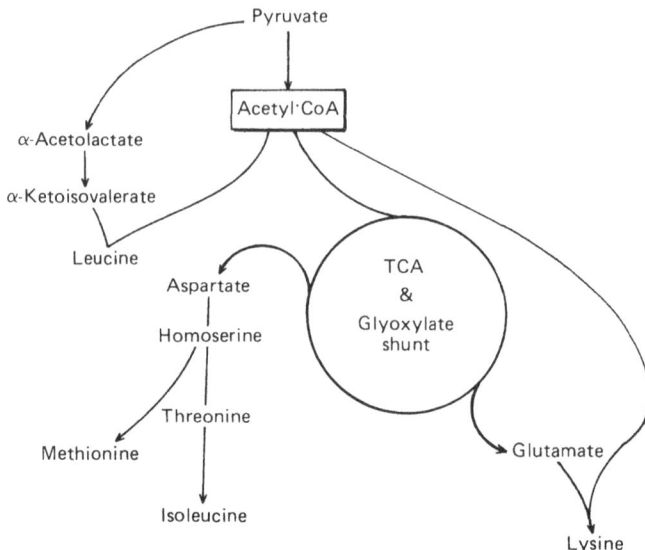

Fig. 12.11. Acetyl coenzyme A and the biosynthesis of amino acids.

acids (Fig. 12.11). If the TCA and the glyoxylate pathway are involved in the biosynthesis of amino acids it is possible to calculate a mass balance which could provide aspartate, glutamate and some energy from acetyl coenzyme A (Fig. 12.12), thus emphasising its key role in amino acid biosynthesis as well as its more obvious roles in fatty acid biosynthesis and energy generation.

An understanding of the control of the availability and use of acetyl coenzyme A must take into account the observation that, in eucaryotes, the TCA and catabolic reactions generating acetyl coenzyme A occur within the mitochondrion, whereas anabolic processes such as fatty acid biosynthesis and secondary metabolite production occur in the cytoplasm (Packer, 1973).

If we consider farnesyl pyrophosphate (Fig. 12.13), which would be the product of the head to tail linkage of three isoprene units derived from three molecules of mevalonic acid, one can see some of the sources of diversity in the sesquiterpenes. There are four possible geometrical isomers involving the two unsymmetrically substituted double bonds. The two sections of —CH$_2$—CH$_2$— in the molecule allow considerable flexibility for folding the molecule into a variety of three-dimensional shapes. Finally there are the possibilities of hydride ions, methyl groups and even larger fragments migrating in response to the generation of a carbonium ion.

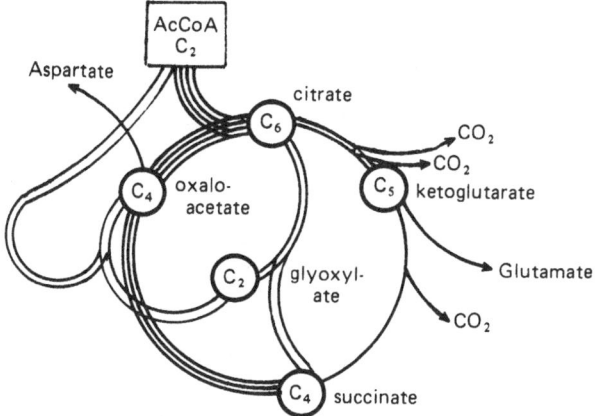

$$6 \text{ Acetyl CoA} + 2 \text{ NH}_3 + 4[\text{H}] \longrightarrow \text{Glutamate} + \text{Aspartate} + 3 \text{ CO}_2 + 18[\text{H}]$$

Fig. 12.12. The production of aspartate, glutamate and energy via the tricarboxylic acid cycle, and the central role of acetyl coenzyme A.

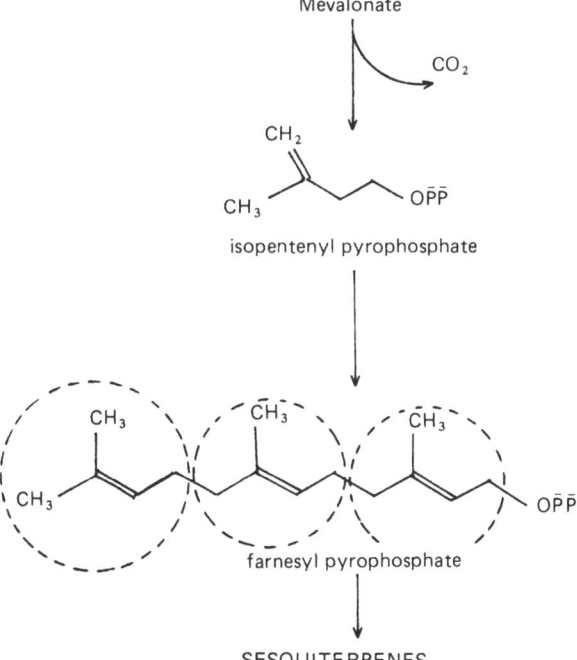

Fig. 12.13. The biosynthesis of farnesyl pyrophosphate, the precursor of the sesquiterpenes including the trichothecenes.

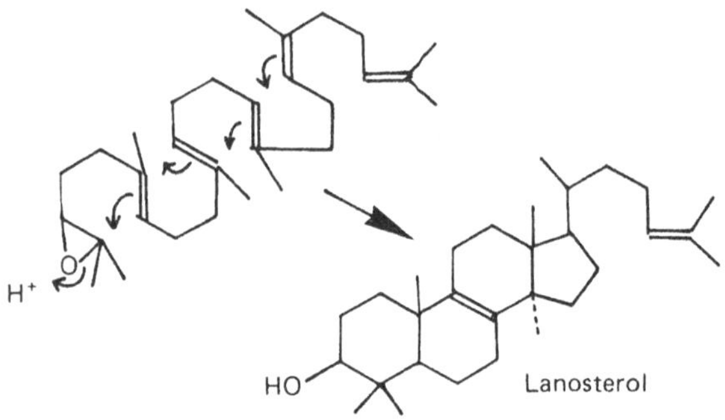

Squalene oxide

Fig. 12.14. Cyclisation of squalene epoxide, an example of the involvement of an epoxide in primary metabolism.

It should be noted that the processes of ring formation, ring contraction, ring expansion, methyl group and hydride ion migration are all part of the biosynthetic processing used for the biosynthesis of sterols. Indeed even epoxide formation is part of the same currency because it is considered that the cyclisation of the squalene chain is initiated by an epoxide group which leaves its mark as the oxygen function so commonly present in the three position of sterols, steroids and triterpenes (Fig. 12.14) (Corey, Russey & Ortiz de Montellano, 1964).

The trichothecenes are considered to be derived from cis-trans-farnesyl pyrophosphate and to involve a double methyl shift (Fig. 12.15) (Hanson & Achilladelis, 1967; Achilladelis, Adams & Hanson, 1970). The compound trichodiene, which was suggested as an intermediate by Bu'Lock (1965), has been isolated as a product of *Trichothecium roseum* Link along with another possible intermediate, trichodiol (Nozoe & Machida, 1972).

Having built up the epoxy trichothecene structure there are several positions around the molecule which can be substituted with oxygen functions. A number of fusaria are known to be able to hydroxylate such compounds as steroids (Marsheck, 1971) and Grove (1969) commented on the remarkable capacity of some fusaria for oxidation of the trichothecene nucleus. Indeed in nivalenol, and such compounds as 7,8-dihyroxy diacetoxy scirpenol, five positions around the molecules are substituted with oxygen (Fig. 12.16). Table 12.2 shows the distribution of the variously hydroxylated trichothecenes amongst a number of species of *Fusarium* as well as the distribution in the production of some other metabolites. This information is based on the author's assessment

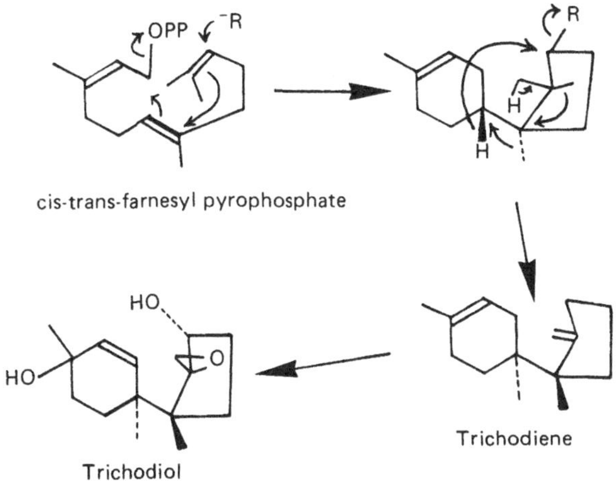

cis-trans-farnesyl pyrophosphate

Trichodiene

Trichodiol

Fig. 12.15. Derivation of trichothecene intermediates from cis-trans-farnesyl pyrophosphate.

Nivalenol

7, 8 Dihydroxy-diacetoxy scirpenol

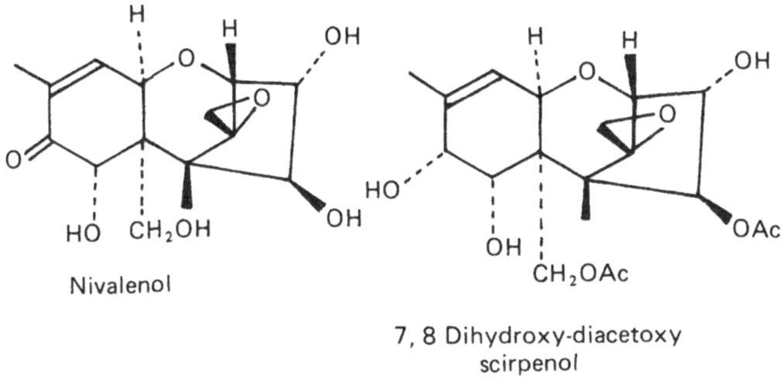

Fig. 12.16. Examples of penta substituted trichothecenes.

of the literature and will probably require revision as further information is confirmed. The alcohol functions are frequently acetylated and it is not always clear in what form the compounds are actually secreted by the mould and how many of the compounds now described in the literature arise from subsequent hydrolysis of O-acetyl derivatives. When one considers that a pentahydric alcohol could give rise to 31 distinct acetates, it is perhaps not surprising that there are so many trichothecene derivatives now recorded in the literature. Grove (1969) had suggested that nivalenol itself may be an artefact of extraction because of the known lability of some of the acetoxy groups. Yoshizawa

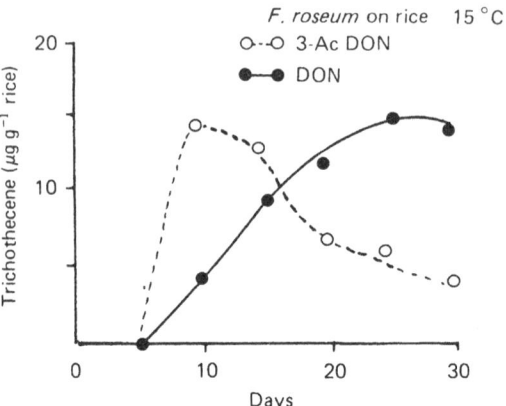

Fig. 12.17. The production of deoxynivalenol (vomitoxin) and its 3-acetyl derivative (redrawn from Yoshizawa & Morooka, 1977).

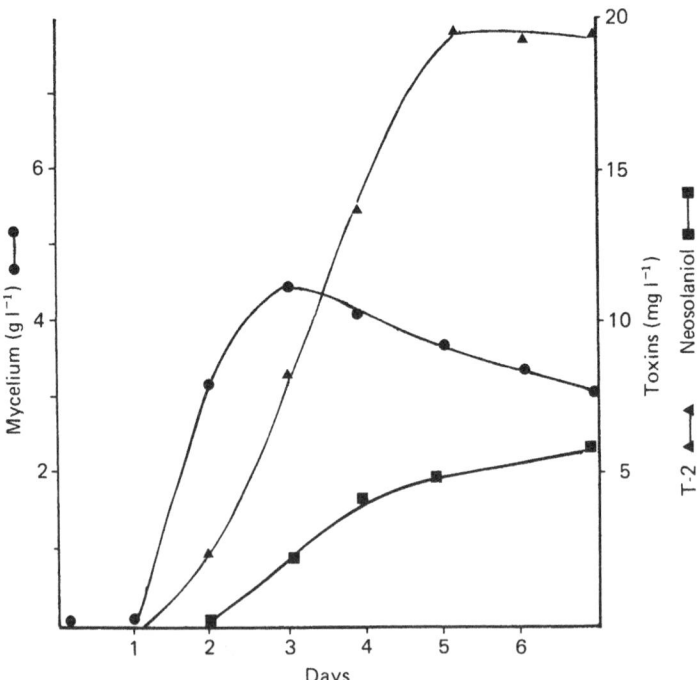

Fig. 12.18. Production of neosolaniol and T-2 toxin by *Fusarium* (redrawn from Ueno, Sawano & Ishii, 1975).

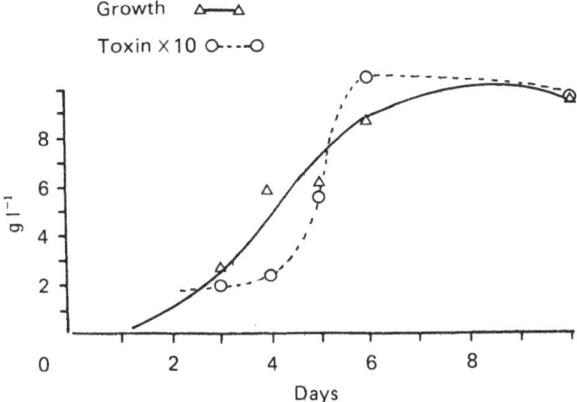

Fig. 12.19. Production of fusarenon-X by *Fusarium nivale* (redrawn from Ueno *et al.*, 1970).

& Morooka (1977) showed that 3-acetyl deoxynivalenol was produced earlier than deoxynivalenol during the growth of *F. roseum* on rice; indeed as the concentration of the latter compound increased that of the former decreased (Fig. 12.17). Ueno *et al.* (1975) showed that neosolaniol may be produced subsequent to the formation of T-2 toxin, the former being a deacylated form of the latter (Fig. 12.18).

Presumably because of the problems of analysis and production in liquid media there are not many records of the relationship between growth and toxin production. From the results of Ueno *et al.* (1970) it can be seen that fusarenon-X production follows the pattern to be expected of a secondary metabolite (Fig. 12.19). It would be interesting to know whether for those strains known to produce both trichothecenes and zearalenone, a similar pattern of production occurs as that demonstrated for bikaverin and gibberellin production by *Gibberella fujikuroi* by Bu'Lock *et al.* (1974).

References

Abrahamsson, S. & Nilsson, B. (1964). Direct determination of the molecular structure of trichodermin. *Proceedings of the Chemical Society*, 118.

Achilladelis, B., Adams, P. M. & Hanson, J. R. (1970). The biosynthesis of the sesquiterpenoid trichothecane antibiotics. *Chemical Communications*, 511.

Arigoni, D. (1975). Stereochemical aspects of sesquiterpene biosynthesis. *Pure and Applied Chemistry*, **41**, 217.

Borrow, A., Jeffreys, E. G., Kessell, R. H. J., Lloyd, E. C., Lloyd, P. B. & Nixon, I. S. (1961). The metabolism of *Gibberella fujikuroi* in stirred culture. *Canadian Journal of Microbiology*, **7**, 227–76.

Brian, P. W., Dawkins, A. W., Grove, J. F., Hemming, H. G., Lowe, D. & Norris, G. L. F. (1960). Phytotoxic compounds produced by *Fusarium equiseti. Journal of Experimental Botany*, **12**, 1–12.

Brian, P. W. & McGowan, J. C. (1946). Biologically active metabolic products of the mould *Metarrhizium glutinosum* S. Pope. *Nature*, **157**, 334.

Bu'Lock, J. D. (1965). *The biosynthesis of natural products*. New York: McGraw Hill.

Bu'Lock, J. D. (1976). Hormones in fungi. In *The Filamentous Fungi*, vol. 2, ed. J. E. Smith & D. R. Berry, pp. 345–68. London: Edward Arnold.

Bu'Lock, J. D. (1980). Mycotoxins as secondary metabolites. In *The Biosynthesis of Mycotoxins, a study in Secondary Metabolism*, ed. P. S. Steyn, pp. 1–16. New York, London: Academic Press.

Bu'Lock, J. D., Detroy, R. W., Hostalek Z. & Munim-Al-Shakarchi, A. (1974). Regulation of secondary biosynthesis in *Gibberella fujikuroi. Transactions of the British Mycological Society*, **62**, 377–89.

Burnett, J. H. (1979). Aspects of the structure and growth of hyphal walls. In *Fungal Walls and Hyphal Growth*, ed. J. H. Burnett & A. P. J. Trinci, pp. 1–25. Cambridge University Press.

Charlwood, B. V. & Banthorpe, D. V. (1978). The biosynthesis of monoterpenes. *Progress in Phytochemistry*, **5**, 65–125.

Cole, R. J., Kirksey, J. W., Cutler, H. G., Doupnik, B. L. & Peckham, J. C. (1973). Toxin from *Fusarium moniliforme*: effects on plants and animals. *Science*, **179**, 1324–6.

Corey, E. J. Russey, W. E. & Ortiz de Montellano, P. R. (1964). 2,3-Oxidosqualene, an intermediate in the biological synthesis of sterols from squalene. *Journal of the American Chemical Society*, **88**, 4750–1.

Delmotte, P. & Delmotte-Plaquee, J. (1953). A new antifungal substance of fungal origin. *Nature*, **171**, 344.

Dimroth, P., Walter, H. & Lynen, F. (1970). Biosynthese von 6-methyl salicylsaure. *European Journal of Biochemistry*, **13**, 98–110.

Freeman, G. G. & Morrison, R. I. (1948). Trichothecin: an antifungal metabolic product of *Trichothecium roseum* Link. *Nature*, **162**, 30.

Gorbik, L. T., Pidoplichko, G. A. & Loiko, Z. I. (1980). Lipids of fungi from *Fusarium* Lk ex Fr. *Mikrobiolomichniyi Zhurnal (Kiev)*, **43**, 191–6.

Grove, J. F. (1968). The constituents of glutinosin. *Journal of the Chemical Society C*, 810–12.

Grove, J. F. (1969). Toxic 12,13 epoxy trichothec-9-enes from *Fusarium* sp. *Chemical Communications*, 1266.

Hagler, W. M. & Mirocha, C. J. (1980). Biosynthesis of (^{14}C) Zearalenone from (1^{14}C) acetate by *Fusarium roseum* Gibbosum. *Applied and Environmental Microbiology*, **39**, 668–70.

Hanson, J. R. & Achilladelis, B. (1967). The role of farnesyl pyrophosphate in the biosynthesis of trichothecin. *Chemistry and Industry*, 1643–4.

Harrison, J. G. (1970). Miscellaneous products from yeasts. In *The Yeasts*, Vol 2, ed. A. H. Rose & J. S. Harrison, pp. 529–45. New York, London: Academic Press.

Inaba, T. & Mirocha, C. J. (1979). Preferential binding of radiolabelled zearalenone to a protein fraction of *Fusarium roseum* 'Graminearum'. *Applied and Environmental Microbiology*, **37**, 80–4.

Jones, E. R. H. & Lowe, G. (1960). The biogenesis of trichothecin. *Journal of the Chemical Society*, pp. 3959–62.

Lynen, F. (1967). Multienzyme complex of fatty acid synthetase. In *Organisational biosynthesis*, ed. H. J. Vogel, J. O. Lampen & V. Bryson, pp. 243–66. New York & London: Academic Press.

McCapra, R., Scott, A. I., Delmotte, P., Delmotte-Plaquee, J. & Bhacca, N. S. (1964). The constitution of monorden, an antibiotic with tranquilising action. *Tetrahedron Letters*, 869–75.

Marsheck, W. J. (1971). Current trends in the microbiological transformation of steroids. *Progress in Industrial Microbiology*, **10**, 49–103.

Mirocha, C. J., Christensen, C. M. & Nelson, G. M. (1971). F-2 (zearalenone) estrogenic mycotoxin from fusaria. In *Microbial Toxins*, vol. 7, ed. S. Kadis, A. Ciegler & S. J. Ajl, pp. 107–38. New York & London: Academic Press.

Mirrington, R. N., Ritchie, E., Shoppee, C. W., Taylor, W. C. & Sternhell, S. (1964). The constitution of radicicol. *Tetrahedron Letters*, pp. 365–70.

Nair, M. S. R. & Carey, S. T. (1980). Metabolites of pyrenomycetes, XII. Structure of (+) hypothemycin, an antibiotic macrolide from *Hypomyces trichothecoides*. *Tetrahedron Letters*, **21**, 2011–12.

Nozoe, S. & Machida, Y. (1972). The structures of trichodiene and trichodiol. *Tetrahedron*, **28**, 5105–11.

Packter, N. M. (1973). *Biosynthesis of Acetate-derived Compounds*. London: John Wiley & Sons.

Ratledge, C. (1982). Microbial oils and fats: An assessment of their commercial potential. *Progress in Industrial Microbiology*, **16**, 119–206.

Robinson, P. M., Park, D. & McClure, W. K. (1969). Observations on induced vacuoles in fungi. *Transactions of the British Mycological Society*, **52**, 447–50.

Springer, J. P., Clardy, J., Cole, R. J., Kirksey, J. W., Hill, R. H., Carlson, R. M. & Isador, J. L. (1974). Structure and synthesis of moniliformin, a novel cyclobutan microbial toxin. *Journal of the American Chemical Society*, **96**, 2267–8.

Steele, J. A., Lieberman, J. R. & Mirocha, C. J. (1974). Biogenesis of zearalenone (F-2) by *Fusarium roseum* 'Graminearum'. *Canadian Journal of Microbiology*, **20**, 531–4.

Tamm, C. L. & Breitenstein, W. (1980). The biosynthesis of trichothecene mycotoxins. In *The Biosynthesis of Mycotoxins*, ed. P. S. Steyn, pp. 69–104. New York: Academic Press.

Ueno, Y., Ishikawa, Y., Saito-Amakai, K. & Tsunoda, H. (1970). Environmental factors influencing the production of fusarenon-X, a cytotoxic mycotoxin of *Fusarium nivale* Fn 20. *Chemical Pharmacology Bulletin*, **18**, 304–12.

Ueno, Y., Sawano, M. & Ishii, K. (1975). Production of trichothecene mycotoxins by *Fusarium* species in shake culture. *Applied and Environmental Microbiology*, **30**, 4–9.

Vesonder, R. F., Tjarks, L. W., Giegler, A., Spencer, G. F. & Wallen, L. L. (1977) 4-acetamido 2-butenoic acid from *Fusarium graminearum*. Phytochemistry, **16**, 1296–7.

Wolf, J. C. & Mirocha, C. J. (1973). Regulation of sexual reproduction in *Gibberella zeae (Fusarium roseum* 'Graminearum') by F-2 (Zearalenone). *Canadian Journal of Microbiology*, **19**, 725–34.

Wolf, J. C. & Mirocha, C. J. (1977). Control of sexual reproduction in *Gibberella zeae (Fusarium roseum* 'Graminearum'). *Applied and Environmental Microbiology*, **33**, 546–50.

Yates, S. G., Tookey, H. L., Ellis, J. J. & Burkhardt, H. J. (1968). Mycotoxins produced by *Fusarium nivale* isolated from tall fescue (*Festuca arundinacea* Schreb.). *Phytochemistry*, **7**, 139–46.

Yoshizawa, T. & Morooka, N. (1977). Trichothecenes from mold-infested cereals in Japan. In *Mycotoxins in Human and Animal Health*, ed. J. V. Rodricks, C. W. Hesseltine & M. A. Mehlman, pp. 309–321. Illinois: Pathotox Publishers Inc.

13
Useful metabolites of *Fusarium*

JOHN D. BU'LOCK

Wolfson Biomass Unit, Weizmann Microbial Chemistry Laboratory, Department of Chemistry, The University of Manchester, Manchester M13 9PL, U.K.

A versatile genus, of ill repute

The genus *Fusarium* would appear to be reasonably versatile in its secondary metabolism, like other filamentous fungi of this kind. The grouping of examples given in Table 13.1 is indicative rather than complete, but it adequately illustrates this versatility. If the examples are taken in detail, and studied more closely, then the major distinctions in biosynthetic patterns which they show should also be found to correspond reasonably closely with taxonomic classifications within *Fusarium*, an aspect which might well deserve sympathetic attention. Also, the two main classes in Table 13.1 each include one of the 'useful' metabolites, by which I mean metabolic products which have been found to have specific biological properties which can be turned to

Table 13.1. *Principal groups of secondary metabolites from* Fusarium *spp*

Class	Subclass	Example	Source
Polyketides	Me,Pr-anthraquinones	Javanicin	*F. javanicum*
	Bu-anthraquinones	Rubrofusarin	*F. culmorum*
	Orsellinyl-anthraquinones	Bikaverin	*F. moniliforme*
	O-alkylbenzoic acid lactones	Zearalenone	*F. graminearum*
Terpenoids	Triprenylphenols	Ascochlorin	*Fusarium* spp.
	Trichothecenes	Nivalenol	*F. nivale*
	Kaurenoids	Gibberellins	*F. moniliforme*
	Fusidanes	Helvolic acid	*F. oxysporum*
Nitrogen compounds	Pyridines	Fusaric acid	*Fusarium* spp.
	Tripeptides	Lycomarasmine	*F. lycopersici*
	Depsipeptides	Enniatins	*Fusarium* spp.

human advantage and profit, and have therefore been manufactured commercially.

Before leaving Table 13.1 it is worth making a further comment. Few if any of the secondary metabolites from *Fusarium* have been discovered as a result of the many searches that have been conducted in screening programmes dedicated to the discovery of useful compounds. Some, and particularly the various quinonoid polyketides which are strikingly coloured, were found through simple curiosity about conspicuously coloured isolates, either as surface colonies or in dispersed cultures; as such (since this type of study began many years ago) they are amongst the best-known examples of fungal pigments. However, the other metabolites were nearly all found through research into the causative agents of toxic effects arising, directly or indirectly, from the growth of *Fusarium* species on plant materials.

In the present book half of the chapters are needed to describe various degrees of nastiness which can be associated with *Fusarium*, and a good deal more than half of the attention which has been given to this genus springs from the same quite justifiable concern. Undoubtedly this has given *Fusarium* a bad reputation, which is much in evidence as soon as any proposals for harnessing it more constructively are made. The most obvious exception to this has of course been the development of very specific strains of *Fusarium* for use in 'mycoprotein' projects, a topic which Dr Solomons discusses fully in the following chapter, and in which the overcoming of this *a priori* verdict has been a very major concern. There has been some recent interest in the ability of fusaria to ferment xylose to ethanol (Suihko & Enari, 1981), but otherwise the genus is equally neglected in general biotechnology, for example as a potential source for useful enzymes. In general, the principle would seem to be 'if you can get it from anywhere else, don't use a *Fusarium*'. This is why the present chapter is specifically concerned with secondary metabolites, which by definition will be relatively unique to the producing strain. Among the secondary metabolites of *Fusarium* there are, as we would expect, products which can not be obtained elsewhere at all. Moreover, what is detrimental in one context can be advantageous in another; one man's mycotoxin is another man's Roquefort!

Gibberellins

'Foolish rice' is the traditional Japanese name for a condition in this staple crop caused by soil-borne infection with the consequently named *Gibberella fujikuroi*, that is, the perfect stage of *Fusarium moniliforme*. Following infection, the leaves and stems of affected seedlings elongate conspicuously before wilting, and as early as 1926 (Kurosawa, 1926) this effect had also been produced with filtrates from cultures of the fungus. An active metabolite, gibberellin A, was first isolated some 12 years later (Yabuta & Sumiki, 1938), but a very large number of gibberellins are now known. There are several reasons for this multiplicity. First, the biosynthesis of gibberellins in *F. moniliforme* has been very intensively studied, so that many intermediates, co-metabolites, and incidental or deliberate metabolic diversions have been characterised. Second, the relatively large number of steps in the later parts of the biosynthetic sequence has the not unusual consequence of generating a large number of such molecular types. Third, the pathway also exists in higher plants where there are several characteristic variations on the pathways most fully explored in the fungus.

The discovery that what could only be considered as a phytotoxic agent from a pathogenic fungus was also a natural plant growth regulator of ubiquitous occurrence and quite fundamental importance in plant physiology has still to be fully assimilated into our views concerning the evolutionary origin of secondary metabolic pathways. With one exception the gibberellins have not been found as products of any species of microorganism other than variously described strains of *F. moniliforme*. The exception is the recent account by Rademacher & Graebe (1979) of the production of gibberellin A4 by the quite unrelated fungus *Sphaceloma manihoticola*, parasitic upon cassava and causing the same kind of pathological elongation as *G. fujikuroi* upon rice. Moreover, despite differences of detail in later parts of the pathway, the essentials of the biosynthesis are the same in the fungus as in plants, with the important difference that even in wild-type *F. moniliforme* the amounts produced are far larger than in any plant tissue. It has never been established whether there is any real homology between the actual enzymes in the fungal and plant systems, such as might for example confirm or exclude their transcription from any DNA sequences of common ancestry; the pursuit of such an hypothesis is now technically possible, and were it to be confirmed, we should have to accept that on at least one occasion in the evolutionary past there had been a wholly illegitimate piece of genetic manipulation, from plant to parasite. It is, of course, the role of gibberellins in plant development that gives these compounds their uses; it is their appearance as secondary metabolites of the fungus that allows those uses to be made practical.

57 varieties

The range of structures in natural gibberellins, already referred to, is actually a very spectacular example of the diversification phase of a biosynthetic pathway. In 1980 Professor Nobutaka Takahashi of the University of Tokyo showed me the structures of no less than 57 free gibberellins from natural sources (Fig. 13.1) plus a dozen or so glucosides and glucose esters. Their numbering (in order of discovery) is quite arbitrary; about half of these gibberellins have been found in the fungus and rather more than half in the various plant systems that have been studied. The number that has been found in both types of source is smaller – perhaps ten – but they include the most important members of the whole series.

Fig. 13.2 outlines the biosynthetic pathway (see below) and gives a basis for our brief discussion of the structural diversities (for position numbering, see the structure of gibberellic acid, at the bottom of the figure). The positions at which natural gibberellins vary are carbons 2, 3, 4, 4a, 5, 6, 7, 8, 8a, and 9 (Fig. 13.1). Nearly all the gibberellins carry a hydroxyl on carbon 2. Apart from this feature the most important variation is at carbon 4a. Initially this carries a methyl group (see the biosynthetic scheme) and this, unchanged or variously oxidised, is retained in the C_{20} gibberellins, of which the key intermediate gibberellin A_{14} is an example. By further oxidation this substituent is replaced by oxygen, thus giving rise to the C_{19} gibberellins; usually this oxygen forms a lactone bridge below ring A with the carboxyl group on carbon 1, as in gibberellic acid itself. Otherwise, the whole gamut of further variations arises through the direct or secondary consequences of oxygenation/dehydrogenation reactions at the remaining positions in the molecule, nearly all of which, indeed, seem to be more or less accessible to this kind of further transformation.

The overall biosynthetic pathways indicated in Fig. 13.2 is now known in very considerable detail (reviews: MacMillan, 1978; Hedden, Mac-Millan & Phinney, 1978). As was first recognised, and demonstrated, by Birch and co-workers (Birch, Rickards & Smith, 1958), the gibberellins are rearranged diterpenes, and the key step initiating their formation is the two-step cyclisation of geranylgeranyl pyrophosphate to *ent*-kaurene. Other fungal diterpenoids result from different cyclisations so that this pattern, first to a copalyl pyrophosphate and thence to *ent*-kaurene, is unique to the series. Similarly unique is the oxidative ring contraction which ensues, generating the central 5-membered ring of the gibberellins proper, as in gibberelin A_{14}. As for the remaining transformations, though a great deal of work has been devoted to the exploration of their precise nature and sequence, it is clear that they do not have the unique significance of the earlier steps mentioned; they conform quite closely to the general pattern of oxidative transformations by both plants and fungi, which are better-known in, but by no means confined to, the steroid series. Their relatively non-specific interplay serves to generate the very large range of products now known, particularly from preparations in which the 'natural' course of things has been perturbed by the experimental set-up.

The commercial production of gibberellins became an objective as soon as their profound effects upon plant growth were realised, and somewhat in advance of our understanding either of the mechanisms of those effects, or indeed of the ways in which they could be brought to bear usefully; indeed, knowledge of the effectiveness of the gibberellins has always been ahead of their utility! The main uses today are horticultural, particularly in fruit-growing (control of premature dropping), in viticulture (yield and quality of grapes, particularly seedless grapes), and to some extent in regularising uniformity and continuity in the malting of barley; however in the latter case there have been regulatory issues, notably in brewing laws under which gibberellic acid becomes an 'additive' and in many countries is automatically prohibited. The annual world market is unlikely to exceed 20 tonnes (so far as I am aware) and after many years of successful operation, actual production in the U.K. has ceased.

The basis of the industrial fermentation was very fully documented by Jefferys (1970) and was recently updated (Vass & Jefferys, 1979) as a kind of obituary. The essential basis is the cultivation of an improved strain of the fungus in a rich nitrogen-limited medium followed by continuous slow feeding of further carbohydrate throughout a prolonged production phase: the eventual limiting factor seems to be the capacity of the equipment to transfer sufficient oxygen to maintain function in a very dense mycelial suspension. Initial product levels (prior to 1961) were probably about 1 g/l, and were eventually raised to perhaps five or more times this level, with appreciable reduction in the duration of the process at the same time. The main commercial product was gibberellic acid, A_3, but there is some demand for other gibberellins for specialised horticultural uses (Turner, 1975).

Accounts of industrial production of gibberellins suggest that the product is normally recovered by extraction of acidified broth filtrates with solvents such as ethyl acetate or methyl isobutyl ketone, followed

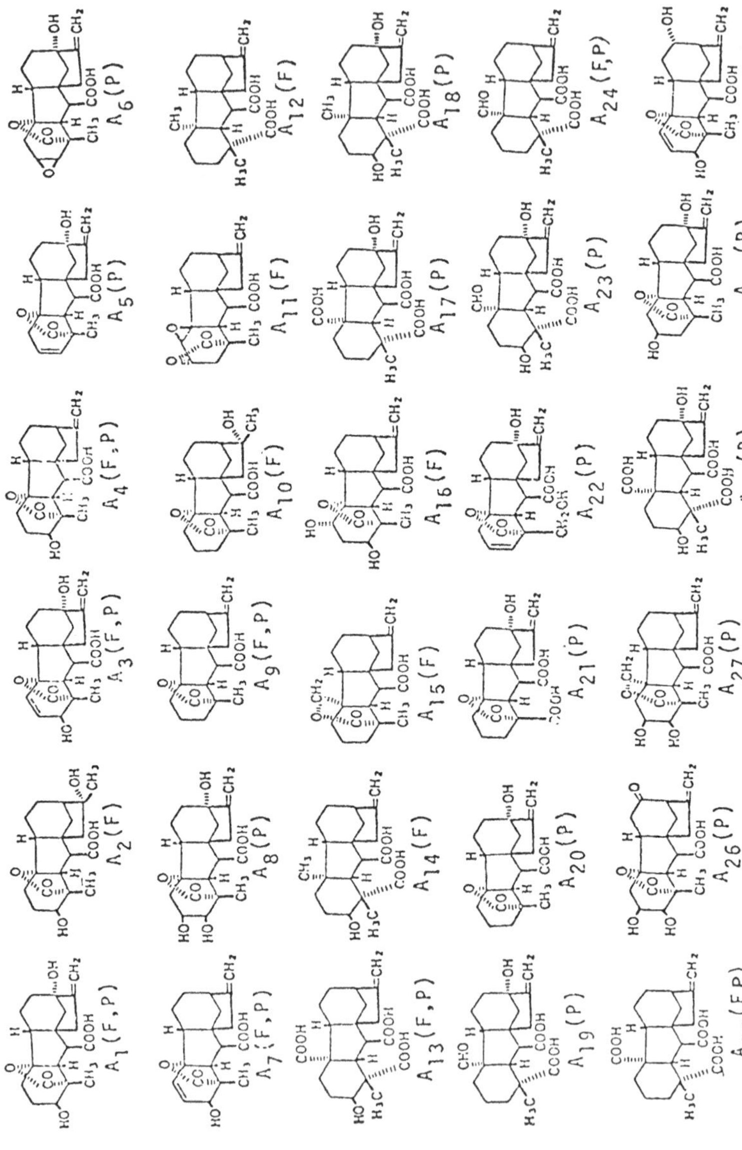
A1 (F,P)
A2 (F)
A3 (F,P)
A4 (F,P)
A5 (P)
A6 (P)
A7 (F,P)
A8 (P)
A9 (F,P)
A10 (F)
A11 (F)
A12 (F)
A13 (F,P)
A14 (F)
A15 (F)
A16 (F)
A17 (P)
A18 (P)
A19 (P)
A20 (P)
A21 (P)
A22 (P)
A23 (P)
A24 (F,P)
A25 (F,P)
A26 (P)
A27 (P)
A28 (P)
A29 (P)
A30 (P)

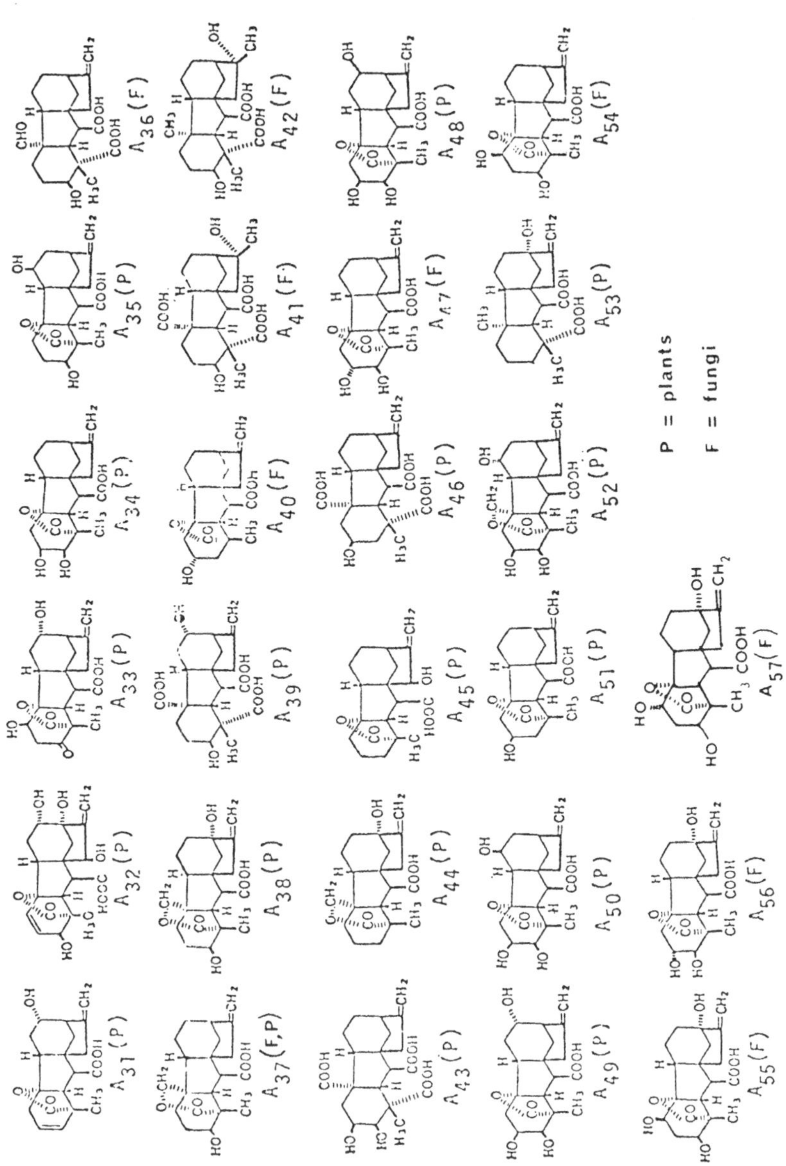

Fig. 13.1. The structures and sources of free gibberellins (from information provided by Professor Nobutaka Takahashi).

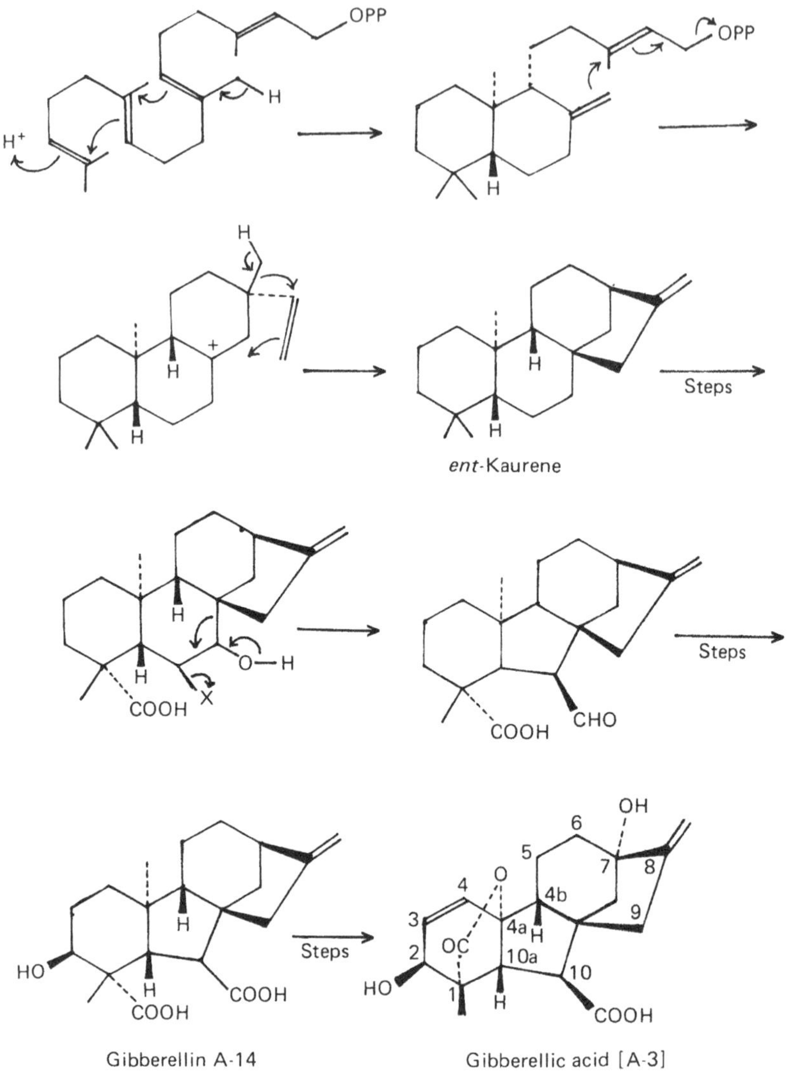

Fig. 13.2. The biosynthesis of gibberellins.

by salt formation with alkali bicarbonates. The fermentation kinetics of gibberellin production have for many years, following the work of the ICI team, provided a classic example of the coupling of growth limitation with secondary metabolite synthesis (Jefferys, 1970 and references there cited). I do not propose to repeat here the often-misquoted generalisations drawn from such instances, but wish to cite only our own study in which the secondary metabolism of *F. monili-forme* was strictly investigated both in batch culture and in relation to growth rates determined by measurable growth limitation in the chemostat; the main observations concerned bikaverin, but data concerning gibberellin production were also given and their implications are equally clear (see below).

Bikaverin

This benzoxanthone quinone (structure, Fig. 13.3) is also a metabolite of *F. moniliforme*, being produced under conditions of less stringent nitrogen limitation than the gibberellins (Bu'Lock *et al.*, 1974). Unlike the gibberellins it is also formed by some other species of *Fusarium* (Balan *et al.*, 1970; Kjaer *et al.*, 1971; Cornforth *et al.*, 1971). Its marked and reasonably specific antiprotozoal activity (Balan *et al.*, 1970) does not seem to have been exploited; presumably because it has some toxic effects which have not been reported. Nor has its vacuolation effect on many fungi, leading to its simultaneous discovery as a 'vacuolation factor' (Robinson, Park & McClure, 1969) been further pursued or explained. From the biosynthetic viewpoint it is such a clear example of a polyketide – albeit with a slightly unusual cyclisation pattern – that it does not appear to have been investigated experimentally except in very preliminary ways. On the other hand the regulation of the biosynthetic pathway in relation to fermentation kinetics was rather fully established in our own work (Bu'Lock *et al.*, 1974) as already noted. The essential conclusions from this study are summarised in Fig. 13.4, in which the upper part shows how in batch culture the initiation, first of bikaverin synthesis, and then of accelerated gibberellin synthesis coincides with progressively more acute limitations in the availability of the nitrogen source (here, glycine, which shows a rather high K_s value so that the later stages of its uptake are 'spread out'). The table in the lower part of Fig. 13.4 confirms this in

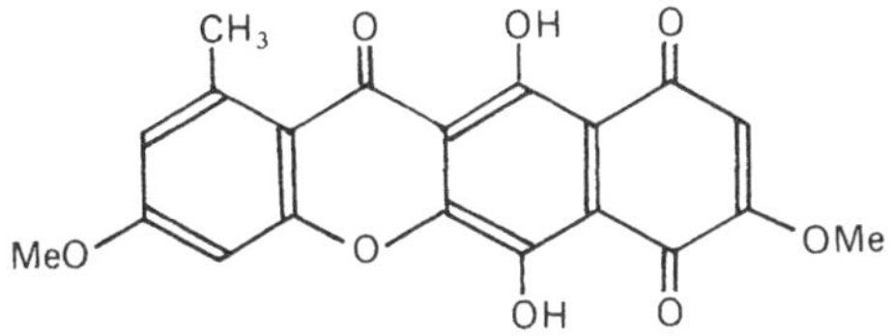

Fig. 13.3. Bikaverin.

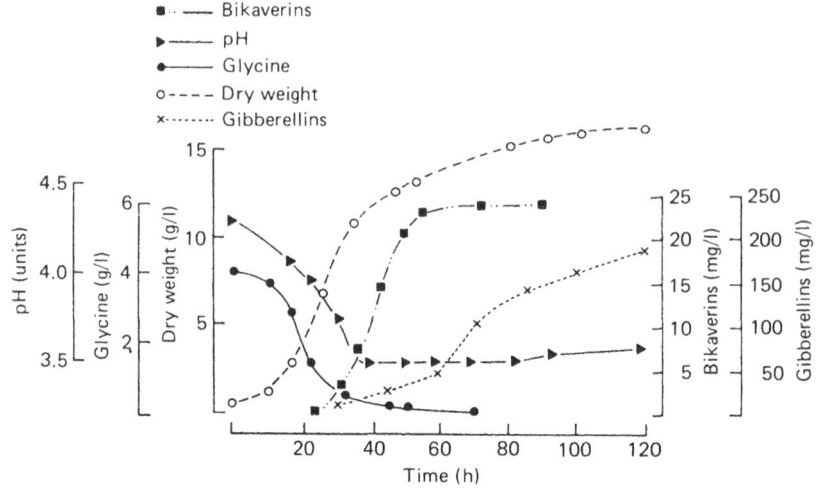

Batch culture data for *G. fujikuroi* on standard medium (limiting glycine).

Specific rates of synthesis in batch and chemostat cultures*

	Batch culture (maximum)	$0.07\ h^{-1}$	Chemostat cultures at		
			$0.05\ h^{-1}$	$0.02\ h^{-1}$	$0.01\ h^{-1}$
Bikaverins	100	10	100	40	n.d.
Gibberellins	300	1	20	40	100

* In μg synthesized per hour per g dry weight of mycelium.

Fig. 13.4. Regulation of bikaverin and gibberellin synthesis in *Gibberella fujikuroi* (Bu'Lock *et al.*, 1974).

terms of progressively lowered dilution rates in glycine-limited chemostat cultures, in which the transient stages seen in batch culture can be indefinitely prolonged. More direct evidence was also provided to show that the level of the enzyme(s) rate-limiting for bikaverin synthesis was controlled directly by the availability of the limiting nutrient and by turnover; the relevant enzyme synthesis could be switched off either specifically by additions of glycine, or by all-round inhibition of protein synthesis with cycloheximide, with wholly equivalent kinetic effects in the two cases.

Whether we shall ever be able to classify bikaverin as a 'useful' metabolite is perhaps doubtful; for example, if it has in fact been found to be too toxic to be a useful antiprotozoal, any prospects for using it as a food colouring agent must be accounted even more dubious – in spite of the fact that the compound was named because the wine-red colour it causes in fermentations recalled a well-known East European vintage! Possibly we can justify its inclusion here as representing 'potentially useful' compounds – a category well-known to writers of grant applications, patent specifications, and similar light fiction.

In the same category, perhaps, we can place some of the strongly cytotoxic *Fusarium* mycotoxins discussed in earlier chapters; every cytotoxic agent provides at least a potential lead into tumour therapy and should surely not be overlooked.

Jekyll or Hyde?

Zearalenone has been frequently referred to in previous chapters of this book. Biosynthetically it is a clear example of a partly-aromatised, partly-reduced nona-ketide; structurally it is an alkyl-resorcylic acid derivative with a 14-membered lactone ring (Fig. 13.5).

It was discovered as the metabolite responsible for vulvovaginitis in female swine fed upon maize infected with the producing fungus, *Fusarium graminearum* (=*Gibberella zeae*) (Stob *et al.*, 1962). From the beginning, however, interest in this particular type of toxic effect (vulvar enlargement, mammary gland stimulation, uterotropic effects, etc.) was combined with the hope that the active principle might have an anabolic or other effect that could be positively exploited in animal nutrition.

Consequently there developed two separate lines of research endeavour, one negative, concerned with the detection, undesirable effects, and avoidance of what was called 'F-2 toxin' in animal feed materials, and one positive, concerned with the production, effectiveness, anabolic activity, regulatory approval and animal tolerance of 'zearalenone' and its derivatives. At one time in the U.S.A., for example, there were two separate programmes thus oriented within the Food and Drugs Administration. The scientific and technical interactions of the two approaches are covered from the manufacturer's

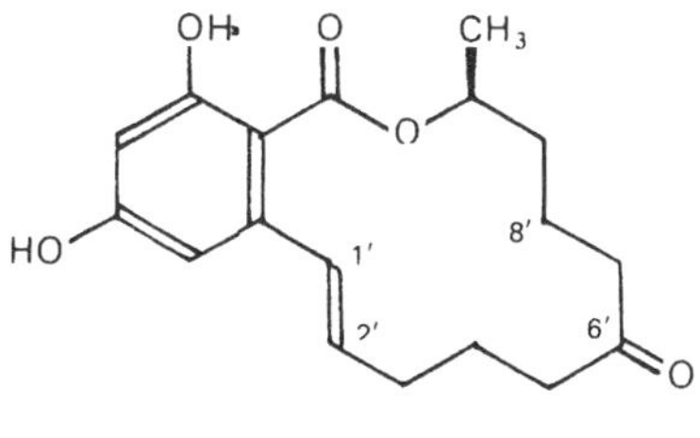

Fig. 13.5. Zearalenone showing positions of important derivatives.

viewpoint (the International Minerals and Chemicals Corporation, formerly Commercial Solvents Corporation) in the very full review of Hidy *et al.* (1977) on which my own account is based. The 'mycotoxin' viewpoint has been more frequently reviewed (e.g. Mirocha, Pathre & Christensen, 1979). Yet a third viewpoint is that in which zearalenone figures as an endogenous regulator of perithecial development in some cultures (Wolff, Lieberman & Mirocha, 1972), but this view has not won acceptance amongst mycologists.

The ketone zearalenone is usually the major product in cultures or casual infections, but the corresponding carbinol (zearalenol) and the 8'-hydroxy-derivatives of both (numbering, see Fig. 13.5) are also found as minor products. Derivatives with the double bond, or both the double bond and carbonyl group, chemically reduced are readily obtained by catalytic hydrogenation of zearalenone and are said to have a better balance of useful properties.

Further transformations of zearalenone in cultures of *F. roseum* seem to involve degradative cleavage of the macrocylic ring (Steele, Mirocha & Pathre, 1976), but in an unidentified fungus what would also seem to be different transformation products of zearalenone (or more strictly of its O-Me derivative) have been found (Ellestad *et al.*, 1978). Interestingly, this last organism, when grown in the presence of ethionine (intended to block the presumed O-methylation), gave not zearalenone but curvularin and its dehydro-derivative; despite any superficial similarity these latter compounds have a different polyketide chain (C_{16} in place of C_{18}) and a quite different cyclisation pattern.

Zearalenone itself has been reported from a range of *Fusarium* species and sub-species, as summarised in Table 13.2 (Mirocha, Pathre & Christensen, 1979). However, even allowing for uncertainties – first of identification, second of nomenclature – it is also quite clear that by

Table 13.2. Fusaria *which may produce zearalenone*

F. graminearum (*=Gibberella zeae*)
F. roseum 'Culmorum'
F. roseum 'Equiseti'
F. roseum 'Gibbosum'
F. roseum 'Graminearum'
F. tricinctum (*=F. sporotrichioides*)
F. oxysporum
F. moniliforme

no means all the isolates which can be ascribed to any one of these designations will have any certainty of producing zearalenone, and the level of any production will also be very variable.

The production of zearalenone by *F. graminearum* and related organisms, whether in infected cereals or in laboratory cultures modelled upon cereal infections, is also subject to a considerable variety of environmental effects; temperature is the most important of these, but also included are nutrient composition and form, air exposure, humidity, inoculum density, light/dark programme, etc. Production of zearalenone on a commercial scale developed on more classic lines – a further example of the intellectual divergence noted earlier – and is fully reviewed by Hidy *et al.* (1977).

Solid-state fermentations were successfully developed first, using increasingly strong glucose-based nutrients (up to 45% sugar!) at 17–18 °C, and giving, after strain improvement which doubled the titre, up to 30 grammes per litre of culture, but only after four to five weeks. Mutant strains which would produce in submerged culture originally gave much lower yields, but the combination of medium improvement, inoculum development, and strain improvement eventually led to fermentations yielding over 30 grammes per litre in under 14 days. Again quite strong sugar nutrient is used (20–30% glucose), and the media have relatively low N and limiting P. The temperature starts around 24 °C and is allowed to rise towards 30 °C; production accelerates as growth slows down. The zearalenone is recovered from the mycelium after filtration, either by extraction with solvent (methanol or acetone) or by extraction into aqueous alkali and precipitation with acid.

Production has been in batches of up to 90 m^3 per vessel; zearalenone derivatives are approved for use as growth promotors in cattle and sheep, at least in the U.S.A., but the scale of the market is not generally known. Nor is the likely future for zearalenone: its toxicity in man is very low, so that even some clinical uses as an oestrogen analogue have been variously advocated – and the residual levels in treated animal carcases are at the lower limit of detection methods. Possibly the increasing trend towards prohibition of more active compounds, such as the true anabolic steroids, will have the effect of increasing the usefulness of zearalenone; equally possible, as part of the underlying trend to prohibit everything (particularly everything being done by a competing nation) its use may suffer. Such is the difficulty of defining a 'useful' product – perhaps especially one from a *Fusarium*!

References

Balan, J., Fuska, J., Kuhr, I. & Kuhrova, V. (1970). Bikaverin, an antibiotic from *Gibberella fujikuroi* active against *Leishmania brasiliensis*. *Folia Microbiologica*, **15**, 479–84.

Birch, A. J., Rickards, R. W. & Smith, H. (1958). The biosynthesis of gibberellic acid. *Proceedings of the Chemical Society*, 1958, 192–3.

Bu'Lock, J. D., Detroy, R. W., Hostalek, Z. & Munim-al-Shakarchi, A. (1974), Regulation of secondary biosynthesis in *Gibberella fujikuroi*. *Transactions of the British Mycological Society*, **63**, 377–89.

Cornforth, J. W., Ryback, G., Robinson, P. M. & Park, D. (1971). Isolation and characterisation of the fungal vacuolation factor bikaverin. *Journal of the Chemical Society* C, 1971, 2786–7.

Ellestad, G. A., Lovell, F. M., Perkinson, N. A., Hargreaves, R. T. & McGahren, W. J. (1978). New zearalenone-related macrolides and isocoumarins from an unidentified fungus. *Journal of Organic Chemistry*, **43**, 2339–43.

Hedden, P., MacMillan, J. & Phinney, B. O. (1978). The metabolism of the gibberellins. *Annual Review of Plant Physiology*, **29**, 149–92.

Hidy, P. H., Baldwin, R. S., Greasham, R. L., Keith, C. L. & McMullen, J. R. (1977), Zearalenone and some derivatives: production and biological activity. *Advances in Applied Microbiology*, **22**, 59–82.

Jefferys, E. G. (1970), The gibberellin fermentation. *Advances in Applied Microbiology*, **13**, 283–323.

Kjaer, D., Kjaer, A., Pederson, C., Bu'Lock, J. D. & Smith, J. R. (1971). Bikaverin and norbikaverin, benzoxanthentrione pigments of *Gibberella fujikuroi*. *Journal of the Chemical Society* C, 1971, 2792–7.

Kurosawa, E. (1926). Experimental studies on the nature of the substance excreted by 'bakanae' fungus. *Transactions of the National History Society of Formosa*, **16**, 213–27.

MacMillan, J. (1978), Gibberellin metabolism. *Pure and Applied Chemistry*, **50**, 995–1004.

Mirocha, C. J., Pathre, S. V. & Christensen, C. M. (1979). Mycotoxins. In *Economic Microbiology III*, ed. A. H. Rose, pp. 488–94. London: Academic Press.

Rademacher, W. & Graebe, J. E. (1979), Gibberellin A4 produced by *Sphaceloma manihoticola*, the cause of superelongation disease of cassava. *Biochemical and Biophysical Research Communications*, **91**, 35–40.

Robinson, P. M., Park, D. & McClure, W. K. (1969). Observations on induced vacuoles in fungi. *Transactions of the British Mycological Society*, **52**, 447–50.

Steel, J. A., Mirocha, C. J. & Pathre, S. V. (1976). Metabolism of zearalenone by *Fusarium roseum 'graminearum'*. *Journal of Agriculture and Food Chemistry*, **24** (i), 89–96.

Stob, M., Baldwin, R. S., Cuite, J., Andrews, F. N. & Gillette, K. G. (1962), Isolation of an anabolic uterotrophic compound from corn infected with *Gibberella zeae*. *Nature*, **196**, 1318.

Suihko, M.-L. & Enari, T.-M. (1981). Production of ethanol from glucose and xylose by different *Fusarium* strains. *Biotechnology Letters*, **3**, 723–8.

Turner, W. B. (1975), Commercially important secondary metabolites. In *The Filamentous Fungi*, vol. I, ed. J. E. Smith and D. R. Berry, pp. 131–133. London: Arnold.

Vass, R. C. & Jefferys, E. G. (1979), Gibberellic acid. In *Economic Microbiology*, vol. III, ed. A. H. Rose, pp. 421–34. London: Academic Press.

Wolff, J. C., Lieberman, J. R. & Mirocha, C. J. (1972). Inhibition of F-2 (zearalenone) biosynthesis and perithecium production in *Fusarium roseum 'graminearum'*. *Phytopathology*, **62**, 937–9.

Yabuta, T. & Sumiki, Y. (1938), The crystallization of gibberellins A and B. *Journal of the Agricultural Chemical Society of Japan*, **14**, 1526.

14
Primary metabolism and biomass production from *Fusarium*

C.ANDERSON AND G.L.SOLOMONS

RHM Research Ltd, The Lord Rank Research Centre, Lincoln Road, High Wycombe, Bucks. HP12 3QR, U.K.

Introduction

Over the past 18 years, RHM Research Ltd has been concerned with the development of a process to produce attractive foods by means of a fermentation using carbohydrate as the substrate and micro-fungi as the organism. Desirable characteristics for the organism and process have been outlined previously (Solomons, 1975) and include protein quality and quantity, lack of toxicity and satisfactory growth parameters. Initially, we carried out experiments with a strain of *Penicillium notatum-chrysogenum*, IMI 138291 (Solomons & Spicer, 1973), but the shortcomings of this organism became unacceptable. Although it had excellent nutritional protein quality, the quantity of protein it contained, 25–28%, was too low to compete with conventional sources of protein. In addition, although it grew well in batch culture, in continuous culture it rapidly overgrew all of the internal surfaces of the fermenter, baffles, sparger and even impeller, so that fermenter performance deteriorated quickly to low levels of productivity due to impaired oxygen transfer capacity. In 1968, we commenced a major screening programme for a more suitable organism, which was carried out both in our own laboratories and in conjunction with the Department of Biochemistry at Imperial College, London under the late Professor E. B. Chain, FRS. The programme lasted for a period of three years, and during that time we examined some 3000 isolates, obtained mainly from soil samples taken on a virtually world-wide basis. During the course of the screen, the target values for protein content changed from not less than 30% on a dry weight basis to not less than 45%. Protein quality requirements, as measured by Net Protein Utilisation (NPU) assay, were not altered from the initial requirement of an

NPU based on true protein equivalent to milk protein, that is, a value of around 65. From the screen, 20 cultures were considered of sufficient potential with regard to protein quantity, amino acid composition, yield value, growth rate and absence of objectionable odour and colour, to grow a 1 Kg batch quantity of each, for initial animal feeding trials. Of these, eight were strains of *Fusarium* and it is interesting to observe that other groups concerned with fungi as a source of protein have also chosen *Fusarium* (Smith, Palmer & Reade, 1975; Macris & Kokke, 1978), which indicates that despite the very wide choice of genera available, *Fusarium* spp. are among the most potentially suitable fungi for use as a protein source.

The organism we chose to investigate in detail was a strain of *Fusarium graminearum* Schwabe, which was subsequently deposited at the Commonwealth Mycological Institute and registered as IMI 145425. During ten years of toxicological trials on both animals and humans, this strain has proved to be non-toxic and of excellent nutritional value and is also non-pathogenic to wheat and maize seedlings. In the next sections, we shall report on aspects of the growth, metabolism and composition of the organism, concluding with a brief description of the process technology.

Growth

To maximise fermenter output, it is essential in production to use continuous culture, and it is in this area that the major part of our effort has been placed. Laboratory studies have also included the use of agar plate, shake flask and batch culture. It is now well established that fungi, given a medium with all the required nutrients, will grow exponentially. By following cell growth in batch culture by either dry weight or optical density measurements, coupled with linear radial extension rate determination and colony appearance on agar plates, we established that the organism had an absolute requirement for biotin and apart from carbohydrate carbon source it did not need any additional organic nutrients (Solomons & Scammell, 1974). Choline chloride increases both the exponential growth rate and the colony extension rate (Table 14.1). Further additions of complex nutrients such as yeast extract or corn steep liquor have no additional effect. Strange & Smith (1971) have shown that choline in wheat anthers stimulates growth of *F. graminearum*. Morphologically, in batch culture, individual hyphal networks have longer interbranch distances when choline

Table 14.1. *Effect of vitamin addition to glucose–mineral medium on growth of* F. graminearum, *IMI 145425, in liquid culture or on agar plates*

Vitamin addition	Exponential growth rate (hr^{-1})	Colony diameter (mm)		
		2 days	3 days	4 days
No addition	No growth			
Biotin 50μg/dm³	0.23	9.3	15.1	26.4
Biotin + choline chloride 100 mg/dm³	0.28	22.0	41.9	62.0

Table 14.2. *The effect of glucose chain length on the* μmax *of* F. graminearum *IMI 145425*

Compound	Number of glucose units	μmax (h^{-1})	td (doubling time) (h)
Glucose	1	0.28	2.48
Maltose	2	0.22	3.15
Maltotriose	3	0.18	3.85

Source: From Anderson *et al.* (1975).

is present in the medium, but in continuous culture, this difference is not apparent.

The absolute maximum temperature of growth is 33 °C with 30 °C being our preferred operating temperature. Over the range of pH 4.5–7.0, there is little effect on growth at 30 °C and our normal operating pH has been set at 6.0.

Carbon sources for growth

The organism can use a wide range of carbon substrates. For the series glucose, maltose, maltotriose, the maximum growth rate falls off with increasing chain length, Table 14.2, and since there is virtually no α-amylase activity, growth on starch is very slow. Sucrose is inverted, but in continuous culture, as the dilution rate approaches the maximum growth rate, the equilibrium fructose concentration rises more rapidly than that of glucose; growth on lactose is poor. Growth occurs with the hexoses mannose and galactose, the pentoses ribose and xylose, and with the organic acids acetate, succinate and lactate, although with the

Table 14.3. *Comparison of cell amino acid and carbohydrate content of F. graminearum, IMI 145425, with glucose or ribose as carbon source*

Carbon source	Amino acids ($mg\ g^{-1}$)	Hexose Carbohydrate ($mg\ g^{-1}$)
Glucose	494	111
Ribose	394	254

latter two, heavy sporulation occurs. The total amino acid content of cells is influenced by carbon source, as illustrated in Table 14.3, comparing the use of glucose and ribose. With ribose there is also a decrease which can mainly be accounted for by an increase in hexose carbohydrate content of the cells.

Uptake of glucose in continuous culture

In order to maintain cell yield and minimise the use of the most expensive substrate, growth in continuous culture is limited by carbon. At steady state, equilibrium is achieved between the growth rate and the external concentration of a limiting nutrient and can be expressed as:

$$D = \mu = \frac{\mu max\ s}{K_s + s}$$

where D = dilution rate, s = substrate concentration, μ = specific growth rate, K_s = saturation constant and μmax = maximum specific growth rate.

We have studied the glucose permease system of our strain by following the initial uptake of radioactively labelled glucose by samples taken from cultures grown continuously in a 3.5 litre laboratory fermenter. The data support the hypothesis that there are two differing permease systems. Similar occurrence of two permeases has been reported for *Neurospora crassa* (Scarborough, 1970). Double reciprocal plots of initial rate of uptake versus concentration are shown in Fig. 14.1. Curve I shows a high affinity system ($K_s \simeq 0.02$ mM); curve II, which is a combination of two sets of data, shows a low affinity system ($K_s \simeq 5$ mM), although trace levels ($< 5\%$ of the total activity) of the high affinity system are present. The development of the high affinity

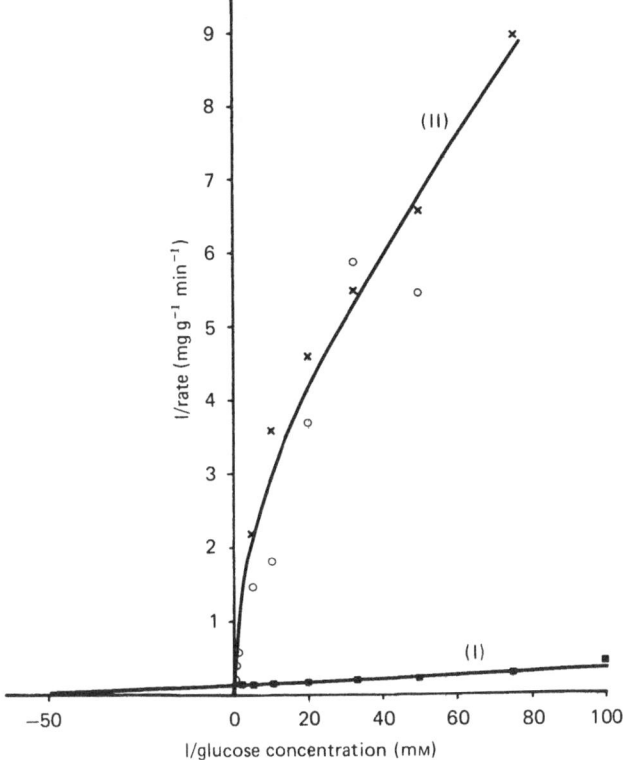

Fig. 14.1 Double reciprocal plots of initial glucose uptake rate versus concentration for *F. graminearum*, IMI 145425.

system can be followed by incubating a sample of mycelium, where the low affinity system predominates, in glucose-free medium and measuring the initial uptake rate of glucose added to sub-samples. The increase in initial uptake rate at 0.05 mM glucose is shown in Fig. 14.2. The presence of two systems theoretically means that at any dilution rate there could be two possible equilibrium carbohydrate concentrations. In practice, only one of the systems would be expected to predominate under any one set of conditions.

Oxygen requirements

Under conditions of submerged culture, the provision of adequate oxygen supply to the cells is technically the most difficult of the growth requirements to fulfil. Since oxygen has such a limited solubility, 7 mg/dm³ at 30 °C at atmospheric pressure, there is only a very limited

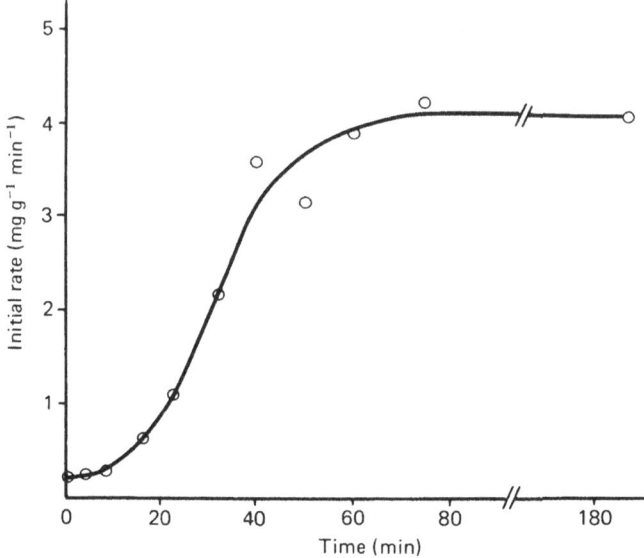

Fig. 14.2 Increase in initial rate of glucose uptake at 0.05 mM versus time following glucose starvation of mycelium of *F. graminearum*, IMI 145425.

reservoir to draw upon. To form a gram of cells of our strain requires 0.78 grams of oxygen and if we are attempting to produce, say 4 kg/m³/h then the oxygen demand is 3.12 Kg/m³/h. Even if the maximum solubility is increased to, say, 14 mg/dm³ by doubling the partial pressure of the system by operating the fermenter at one atmosphere of over-pressure, the oxygen supply in solution would be depleted in about 15 seconds. It is therefore essential to make adequate provision for a continuous supply of oxygen in solution by a suitable means of aeration and agitation. The provision of these parameters is essentially an exercise in chemical engineering, but its importance is difficult to exaggerate. Our approach to this problem has been the subject of a recent publication (Anderson, LeGrys & Solomons, 1982).

The consequences of failing to meet the organism's oxygen requirement are two-fold, although interrelated. Firstly, under oxygen limitation, the cells divert a portion of the carbon into the formation of by-products such as ethanol, a common occurrence with *Fusarium* (Griffin, 1981). This diversion of metabolism is highly undesirable since it can trigger the formation of substantial numbers of macroconidia as well as causing the formation of other compounds with undesirable olfactory properties. Secondly, the yield factor based on carbon utilisation falls and this has severe economic penalties and hence must be avoided.

Nitrogen source for growth

For economic reasons, it is essential that an organism intended for protein production be able to utilise inorganic forms of nitrogen since these are, on a nitrogen basis, much cheaper than complex forms, such as soya meal. Therefore, we use ammonia, which is in fact the ideal nitrogen source for protein, since it can be supplied on demand as a gaseous form linked through a pH control system. Due to its high solubility in water, it presents no problems in mass transfer resistance and levels of ammonium ion in solution can be kept at any desired level. The organism can also utilise nitrate and urea as a source of nitrogen. Because the whole object of growing organisms for protein production is so clearly concerned with protein content, it is surprising to find in the literature culture medium with high C/N ratios of 10/1 or even 20/1 (Gray, 1964). Our organism will normally contain around 9% of nitrogen and 45% carbon, so we always utilise a C/N ratio less than 5/1 to ensure that nitrogen is not the limiting factor in protein production.

Sulphur metabolism

As in most proteins of microbial origin, the sulphur amino acids, especially methionine, limit the nutritional value of the material. Hence, we were interested to examine the sulphur metabolism of the organism to find if there were constraints of sulphur availability that limited methionine formation. It was our view that it would be unlikely that substantial changes of methionine content would be brought about since many different protein species would have to be influenced for that to occur; this view was also confirmed by the work of Alroy & Tannenbaum (1977).

Our organism will use a range of sulphur sources, both inorganic, e.g. sulphate, thiosulphate and tetrathionate, and organic such as methionine, cysteine and cysteic acid. In batch culture using S^{35}-sulphate, mycelium has been analysed for sulphur distribution. In these experiments, the sulphur pool ranged from 1.1–1.8 mg S/g, which represented 30–38% of the total sulphur content. This pool is used as a sulphur reserve and analysis has indicated that the major constituent is taurine. The compound is stable to performic acid oxidation, hydrochloric acid hydrolysis and is ninhydrin positive. Using thin-layer chromatography and paper electrophoresis it co-chromatographed in five

Table 14.4. *Relative taurine content during batch growth of F. graminearum, IMI 145425, with various sulphur sources in glucose–mineral medium + biotin*

Sulphur source	Relative taurine content
Sulphate	100
Thiosulphate	<2
L-Methionine	0
L-Cysteine	0
L-Methionine + Sulphate	33
L-Cysteine + Sulphate	100

Note: Taurine content on sulphate was in the range 5.1–6.4 mg g^{-1}.

solvents and at three pH's with authentic taurine. The presence of taurine is unusual since choline-O-sulphate is normally found in many microorganisms and is believed to be the sulphur reserve (Wright & Vining, 1976). Separate determination using ion-exchange chromatography, based on the method of Moorhouse, Law & Maddix (1976), showed the pool content of taurine, with sulphate as sulphur source, to vary up to 7 mg/g depending on growth conditions. The relative pool taurine content of mycelium during growth in batch culture with various sulphur sources is shown in Table 14.4.

The sensitivity of the organism to amino acid analogues of methionine, particularly ethionine, was investigated as part of a mutation programme initially intended to yield methionine overproducing strains. The relative effect on exponential growth rate in batch culture on minimal medium is shown in Table 14.5. α-Methyl-DL-methionine and DL-norleucine have comparatively little effect. DL-Ethionine has a more marked effect, but germination and some growth is still evident on agar plates at 18 g/dm^3; germination time is also delayed. The organism is therefore resistant to ethionine; by contrast, the growth of *Neurospora crassa* is completely inhibited at 50 mg/dm^3 of L-ethionine (Metzenberg, 1968).

The mechanism of resistance to ethionine is by detoxification. Using radioactively labelled methionine or L-ethionine as substrate, added at 0.5 mM to cells from a continuous culture resuspended in fresh medium

Table 14.5. *Relative effect of antimetabolite concentration on the exponential growth rate of* F. graminearum, *IMI 145425, compared to control (= 100)*

Antimetabolite	Concentration (g dm^{-3})		
	0.1	1.0	10
α-Methyl-DL-methionine	100	100	ND
DL-Norleucine	88	88	76
DL-Ethionine	76	60	52

Note: ND = not determined.

without added carbohydrate, the uptake was followed with time. The net uptake of labelled compound is shown in Fig. 14.3; after 30–40 minutes net excretion of labelled compounds is evident. Additional experiments have shown that the product from methionine is α-keto-γ-methiolbutyric acid (αKM), i.e. deaminated methionine. It forms a precipitate with 2,4-dinitrophenyl hydrazine (as does the ethionine excretion product) and using thin-layer chromatography it has identical Rf values in three solvents with authentic αKM. An assay to determine αKM was developed from that of Fields & Dixon (1971) for reactive carbonyl groups and the appearance of αKM in the medium after addition of L-methionine at 0.5 mM to an aerated cell suspension was followed. Cycloheximide (actidione) at 100 μg/ml has a significant effect on the time course as shown in Fig. 14.4. Major excretion of a αKM is prevented and there is no net sulphur excretion; it was not established whether accumulated material in the cell remained mainly as methionine. Comparing addition of methionine, valine, threonine, serine, phenylalanine and lysine to mycelium showed that only for methionine was there net excretion and substantial keto product after a 90 minute incubation period. This indicates a relatively specific response of the organism to methionine and ethionine.

Initial uptake rates of methionine transport of the organism have been studied on samples of mycelium grown under a number of conditions. Mycelium has been grown under batch and continuous culture with limited and excess glucose and with the addition of DL-methionine or choline during growth, in various, but not all, combinations. Double reciprocal plots of initial rate versus substrate

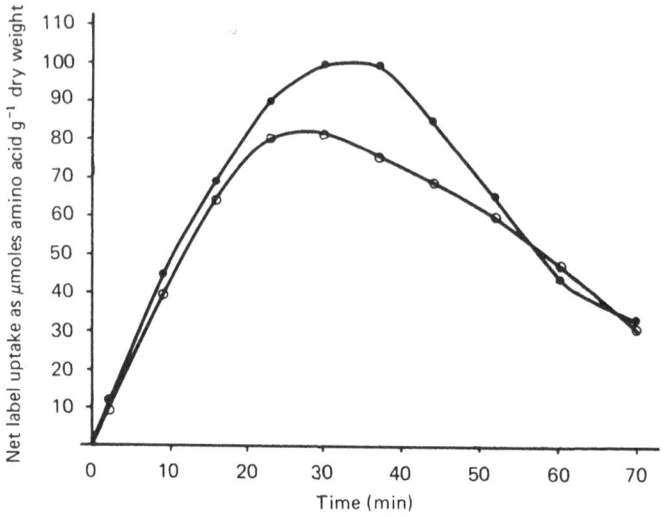

Fig. 14.3. Time course of net uptake of L-methionine (methyl-C[14]) (●) and L-ethionine (ethyl-l-C[14]) (○) for *F. graminearum*, IMI 145425.

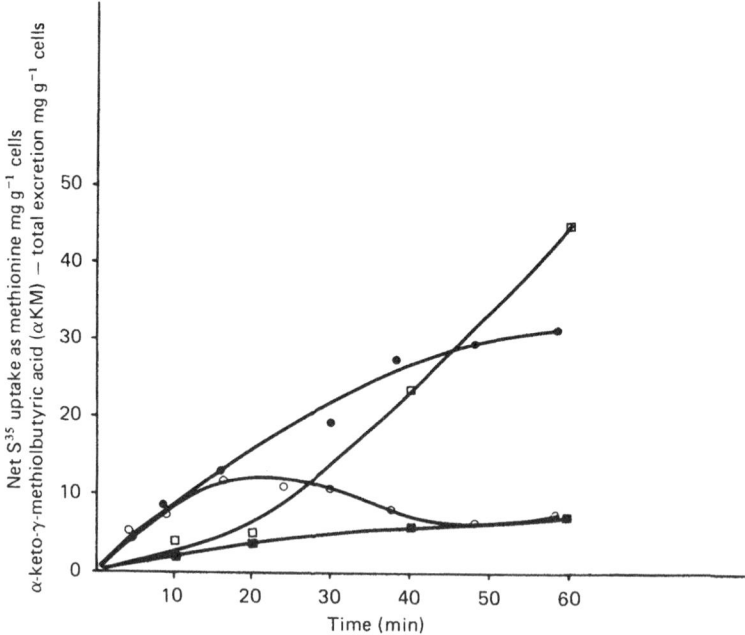

Fig. 14.4. Effect of cycloheximide on net L-methionine S[35] and α-keto-γ-methiolbutyric acid (αKM) formation by *F. graminearum*, IMI 145425. ○ cells; ● cells + cycloheximide; □ α KM excreted; ■ α KM excreted + cycloheximide.

Table 14.6. *Range of kinetic parameters of methionine permeases of* F. graminearum, *CMI 145425, found under various growth conditions*

	K_s (mM)	V_m (mg g^{-1}min^{-1})
System I	0.2–0.5	0.6–2.7
System II	0.007–0.018	0.12–0.93

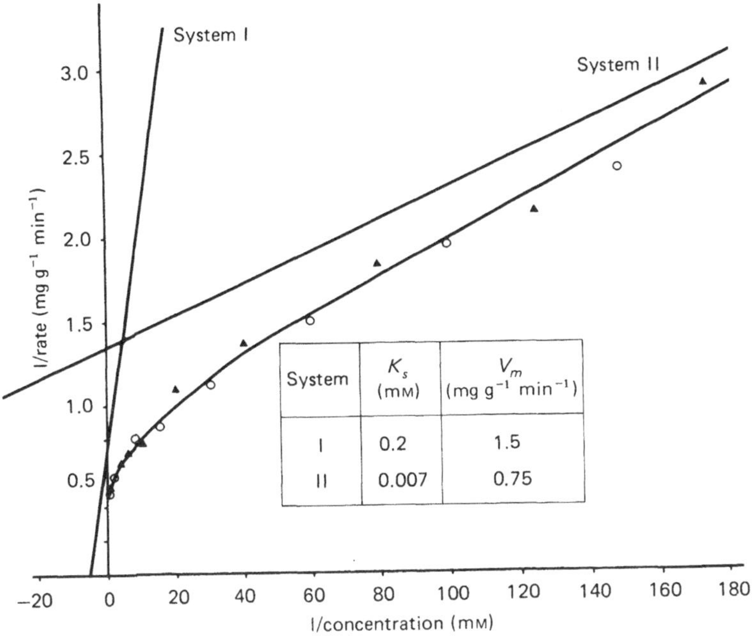

System	K_s (mM)	V_m (mg g^{-1} min^{-1})
I	0.2	1.5
II	0.007	0.75

Fig. 14.5. Example of double reciprocal plot of initial rate versus concentration for methionine uptake by *F. graminearum*, CMI 145425. ○, ▲, experimental points. The straight lines are the theoretical plots, using the constants shown in the insert, for the two derived uptake systems.

concentration are usually non-linear and this has been interpreted as the presence of two permeases with differing affinity constants: an example is shown in Fig. 14.5. Multi-permease systems for amino acids in fungi are well-documented (Whitaker, 1976), although only a single system for methionine in *Fusarium oxysporum* f. sp. *lycopersici* has been reported (Barran, 1981). The range of kinetic constants found is shown in Table 14.6. The maximum rate is reduced when uptake studies are carried out under nitrogen and also in the presence of glucose when

Table 14.7. *Proximate analysis and nitrogen distribution of* F. graminearum, *IMI 145425, (a) before, and (b) after RNA reduction*

Proximate analysis	(a)	(b)
Protein (α-AAN $\times$ 6.22)	42.4	44.3
RNA	9.7	1.1
Lipid	13.0	13.8
Ash	6.6	3.1
Fibre	12.7	18.3
Carbohydrate (by difference)	15.6	19.4
Total	100.0	100.0
Nitrogen distribution		
Free amino acid-N	0.7	—
Protein-N	6.3	7.1
Nucleotide-N	0.2	—
RNA-N	1.5	0.1
n-Acetyl glucosamine-N	1.0	1.5
Total	9.7	8.7
TN (by analysis)	9.7	8.8

growth conditions are such that the high affinity glucose permease system would be expected to predominate. The higher affinity values are also obtained with mycelium grown in continuous culture with low external glucose concentrations.

Cell composition

The composition of fungal cells varies markedly between different genera, species and even strains. We have isolated many fungi with a total nitrogen content of 3–4% whereas occasionally isolates provide values in excess of ten per cent. We have previously reported on a range of protein compositions for twenty organisms (Anderson *et al.*, 1975). When grown in a $1.3\,m^3$ continuous culture vessel with carbon limitation at growth rates between $0.15–0.20\,h^{-1}$, our strain of *Fusarium graminearum* has a proximate analysis as shown in Table 14.7. The composition is unremarkable except for the high protein content. The amino acid analysis of this protein is shown in Table 14.8. It should be noted that the protein:RNA ratio of fresh cells is approximately 4:1

Table 14.8. *Amino acid composition of F.* graminearum, *IMI 145425, (a) before and (b) after RNA reduction (g 100 g protein⁻¹)*

Amino acid	(a)	(b)
Lysine	7.5	7.6
Methionine	1.9	2.2
Cystine	0.6	0.7
Threonine	5.3	5.5
Tryptophan	2.2	1.5
Valine	6.1	5.9
Leucine	7.6	8.1
Iso-leucine	4.2	4.9
Phenylalanine	4.4	4.9
Histidine	2.6	2.7
Total sulphur	(2.5)	(2.9)
Arginine	6.6	6.7
Tyrosine	3.7	3.9
Aspartic acid	10.2	9.8
Serine	5.4	5.6
Glutamic acid	14.4	12.8
Proline	5.4	5.5
Glycine	5.2	5.1
Alanine	7.7	6.7
	101.0	100.1

and the DNA content is low at $< 0.5\%$. The lipid, 14%, extracted with 3:1 chloroform:methanol, occurs mainly as phopholipids; extraction with petroleum ether yields only around 25% of this value. The cell walls are composed largely of β-glucans and chitin, as has been previously reported for fungi (Burnett, 1968). Photomicrographs of the cells (Figs. 14.6, 14.7) illustrate the well-defined cell walls and packed ribosome content of the cells.

RNA reduction

If a microbial source of protein is to be utilised for human food, then particular attention has to be paid to the RNA content of the material. It was recommended by the Protein Advisory Group of the WHO/FAO (1975) that the ingestion of additional RNA from a novel protein source should be limited to 2 g/day. As our strain has an RNA content of 8–9% ex-fermenter, this would limit the ingestion of the material to not more than 20 g/day. Clearly, this would greatly constrain

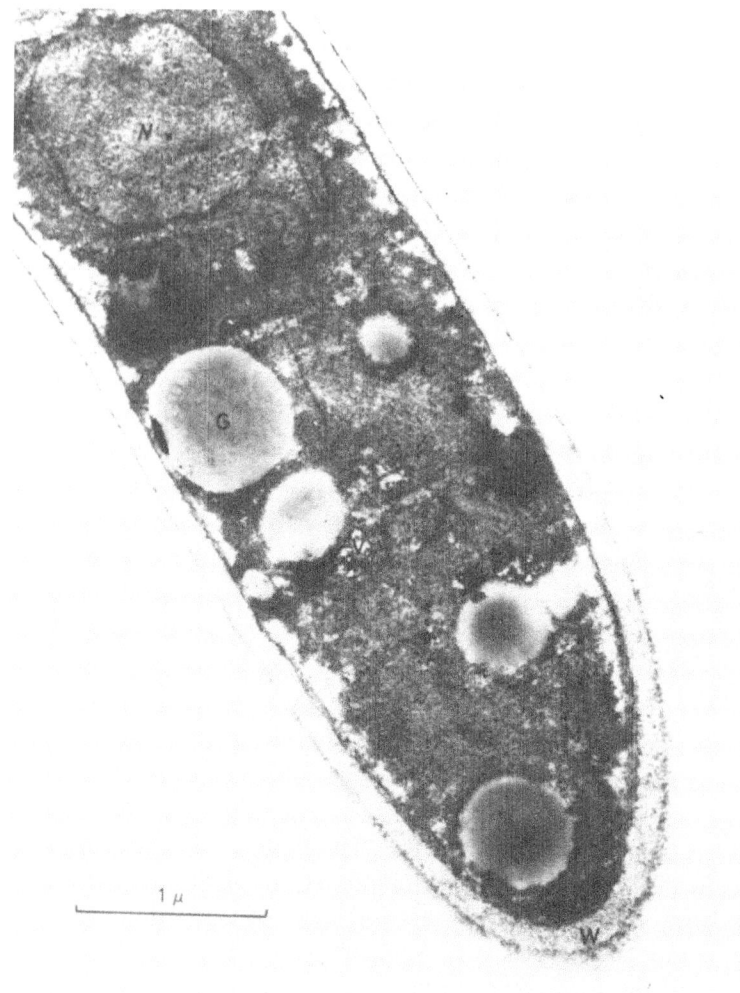

Fig. 14.6. Longitudinal section of *F. graminearum* grown in submerged culture. N = nucleus; G = glycogen, M = mitochondria; V = vacuoli; W = wall.

the potential use of the material and consequently methods have been developed to reduce the content of RNA to levels that would not, in practical terms, impose restrictions on its utilisation.

Initially, we developed a method based upon the treatment of cells with a 2 or 3 carbon alcohol followed by incubation in buffer for periods of 10–30 minutes (Towersey, Longton & Cockram, 1975). The efficien-

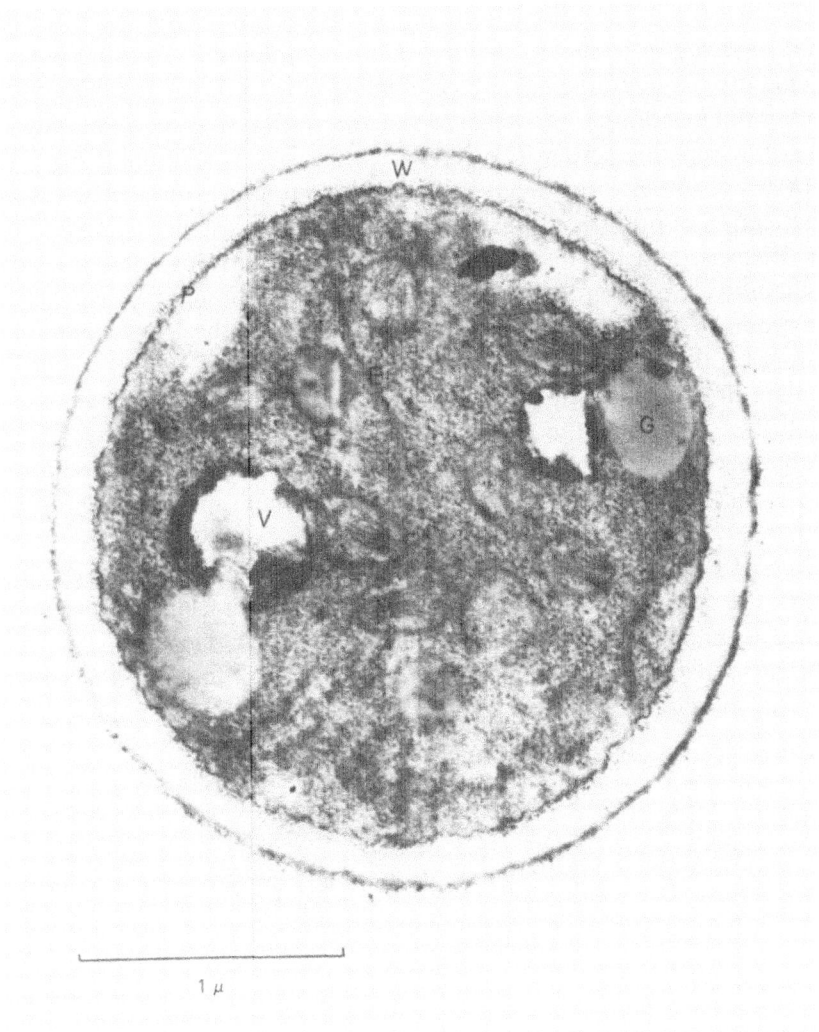

Fig. 14.7 Cross section of *F. graminearum* grown in submerged culture.
W = wall; V = vacuole; G = glycogen.

cy of the system was dramatic, sometimes reducing the RNA to levels
that were undetectable. However, the use of solvents was undesirable
due to their cost and to the inherent inadvisability of using solvent
systems in processing human foods. We therefore re-examined in detail
the heat shock system devised by Maul, Sinskey & Tannenbaum (1970).
Their original method consisted of a three stage treatment and took

about two hours for completion. In our adaptation of the method, we finally arrived at a single stage treatment completed in 20–30 minutes, having the considerable advantage of being able to be carried out under isothermal conditions in a stirred-reactor rather than necessitating a plug-flow pipeline system (Towersey, Longton & Cockram, 1976).

In its essentials, the heat shock treatment relies on the selective inactivation of proteases, due to their lower heat resistance compared to the RNases, and the disruption of ribosomes, allowing for the ready availability of ribosomal RNA for enzyme degradation. After hydrolysis the products, mostly 5′ nucleotides, diffuse through the cell wall into the culture broth. The consequences of RNA reduction on overall composition are profound. The proximate analysis after this process is shown in Table 14.7, although the amino acid composition is much less affected (Table 14.8). The key point is that the protein/RNA ratio is altered from 4:1 to 40:1, thus removing the restrictions on consumption mentioned earlier.

Recently, we have re-investigated the reduction process to develop a model for the reaction. Cells grown in continuous culture were subjected to batch isothermal RNA reduction at 64 °C and pH 6.0 in the laboratory at a known concentration – around 20 g/dm³ – in the culture broth. The reduction, which is illustrated in Fig. 14.8, shows the level of

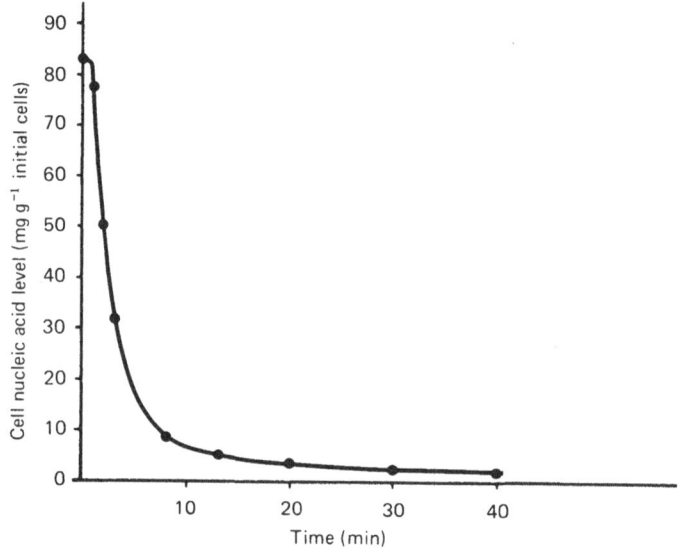

Fig. 14.8. Time course of nucleic acid reduction for *F. graminearum*, IMI 145425.

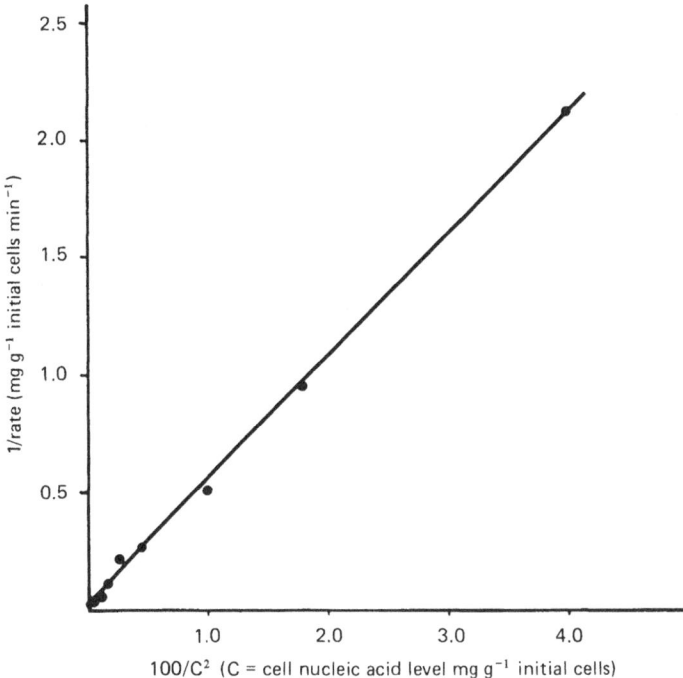

Fig. 14.9. Double reciprocal plot of the rate of nucleic acid removal versus the square of the cell nucleic acid level for *F. graminearum*, IMI 145425.

nucleic acid per initial dry weight unit versus time. This method of plotting approximately relates the decrease to a fixed volume of cells. There is an accompanying loss of dry weight as additionally other low molecular weight compounds are lost and therefore the actual content as mg/g dry weight, at any time, is higher than shown. The rate of disappearance was calculated from the gradient, and the double recip-rocal plot of rate of removal versus the square of the cell level of nucleic acid is shown in Fig. 14.9.

The linear correlation is highly significant ($P < 0.001$) and therefore the relationship can be expressed in the form:

$$V = \frac{V_m C^2}{K + C^2}$$

where V = rate of nucleic acid removal, mg/g/min, V_m = maximum rate, K = constant and C = nucleic acid, mg/g, in initial cells. This strongly

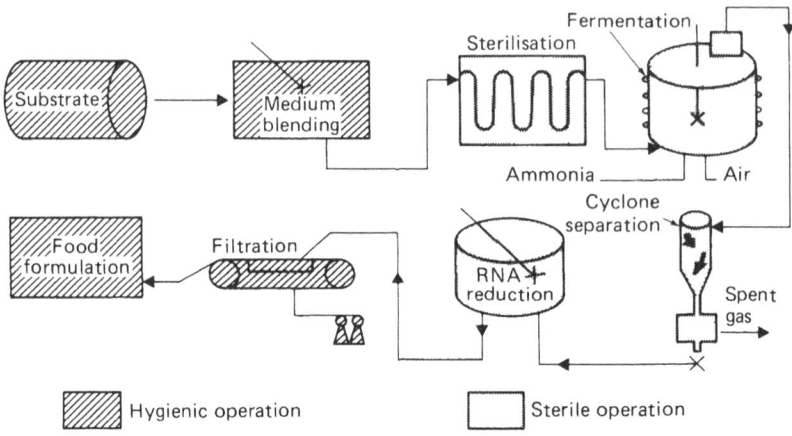

Fig. 14.10. Mycoprotein process flow diagram.

indicates an enzyme reaction where there is activation by the substrate (Mahler & Cordes, 1966).

Process technology

A process flow diagram is shown in Fig. 14.10. Culture medium, sterilised by continuous high temperature–short residence treatment, is fed to the fermenter at a flow giving a dilution rate in the range $0.17–0.20\,h^{-1}$. Nitrogen is supplied as ammonia in the inlet air stream and controlled by pH demand. Agitation is by means of a novel dual-impeller system, which separates and optimises the function of bulk flow and mass-transfer (Solomons & LeGrys, 1981). The product stream is separated from the exhaust air by a cyclone and immediately passed to a stirred tank reactor, with a mean residence time of 20–30 minutes, to reduce the RNA content. Both the fermenter and the RNA reduction stages are strictly aseptic. From the RNA reduction stage, the product stream emerges into a non-sterile but hygienic environment for recovery by vacuum filtration. The output of this filter yields a product termed 'mycoprotein' and consists of a 30% solids filter cake. From this raw material, food technology takes over and can produce a wide range of interesting and nutritious end products.

Acknowledgements. We wish to thank Dr R. E. Angold of the LRRC Microscopy Unit for the photomicrographs in Figures 14.6 and 14.7, and MIT Press for permission to use Table 14.2.

References

Alroy, Y. & Tannenbaum, S. R. (1977). Phenotypic modification in amino acid profiles of cell residues of *Candida utilis* and *Enterobacter aerogenes. Biotechnology and Bioengineering*, **19**, 1155–69.

Anderson, C., LeGrys, G. A. & Solomons, G. L. (1982). Concepts in the design of large-scale fermenters for viscous culture broths. *The Chemical Engineer*, **No. 377**, Feb. 43–9.

Anderson, C., Longton, J., Maddix, C., Scammell, G. W. & Solomons, G. L. (1975). The growth of microfungi on carbohydrates. In *Single-Cell Protein II*, ed. S. R. Tannenbaum, & D. I. C. Wang, pp. 314–29. Cambridge, Mass: MIT Press.

Barran, L. R. (1981). Methionine transport by mycelia of *Fusarium oxysporum f.* sp. *lycopersici. Canadian Journal of Microbiology*, **27**, 743–7.

Burnett, J. H. (1968), *Fundamentals of Mycology*, pp. 19–23. London: Edward Arnold Ltd.

Fields, R. & Dixon, H. B. F. (1971). Micro method for determination of reactive carbonyl groups in proteins and peptides, using 2,4-dinitrophenylhydrazine. *The Biochemical Journal*, **121**, 587–9

Gray, W. D. (1964) Process of the production of fungal protein. US Patent 3151038.

Griffin, D. H. (1981). *Fungal Physiology*, p. 81. New York: John Wiley & Sons.

Macris, B. J. & Kokke, R. (1978). Continuous fermentation to produce fungal protein. Effect of growth rate on the biomass yield and chemical composition of *Fusarium moniliforme. Biotechnology and Bioengineering*, **20**, 1027–35.

Mahler, H. R. & Cordes, E. H. (1966). *Biological Chemistry*, pp. 258–61. New York: Harper & Row.

Maul, S. B., Sinskey, A. J. & Tannenbaum, S. R. (1970). New process for reducing the nucleic acid content of yeast. *Nature, Lond.*, **228**, 181.

Metzenberg, R. L. (1968). Repair of multiple defects of a regulatory mutant of *Neurospora* by high osmotic pressure and by reversion. *Archives of Biochemistry and Biophysics*, **125**, 532–41.

Moorhouse, C. O., Law, A. R. & Maddix, C. (1976). Automated methods for the determination of total amino acids, cystine and methionine in microbial biomass. *Technicon International Congress: Advances in Automated Analysis*, **2**, 182–9.

PAG (1975). *PAG ad hoc* working group meeting on clinical evaluation and acceptable nucleic acid levels of single cell protein for human consumption. (Report of the) *PAG Bulletin*, **5**(3), 17–26.

Scarborough, G. A. (1970). Sugar transport in *Neurospora crassa*. II. A second glucose transport system. *Journal of Biological Chemistry*, **245**, 3985–7.

Smith, R. H., Palmer, R. & Reade, A. E. (1975). A chemical and biological assessment of *Aspergillus oryzae* and other filamentous fungi as protein sources for simple stomached animals. *Journal of the Science of Food and Agriculture*, **26**, 785–95.

Solomons, G. L. (1975). Submerged culture production of mycelial biomass. In *The Filamentous Fungi*, vol. 1, ed. J. E. Smith & D. R. Berry, pp. 249–64. London: Edward Arnold.

Solomons, G. L. & LeGrys, G. A. (1981). British Patent, 1584103.

Solomons, G. L. & Scammell, G. W. (1974). Improvements in or relating to microorganisms. British Patent 1346061.

Solomons, G. L. & Spicer, A. (1973). Improvements in the production of edible protein substances. British Patent 1331471.

Strange, R. N. & Smith, H. (1971). A fungal growth stimulant in anthers which predisposes wheat to attack by *Fusarium graminearum. Physiological Plant Pathology*, **1,** 141–50.

Towersey, P. J., Longton, J. & Cockram, G. N. (1975). Improvements in or relating to the production of edible protein containing substances. British Patent 1408845.

Towersey, P. J., Longton, J. & Cockram, G. N. (1976). Improvements in or relating to the production of edible protein containing substances. British Patent 1440642.

Whitaker, A. (1976). Amino acid transport into fungi: An essay. *Transactions of the British Mycological Society*, **67,** 365–76.

Wright, J. L. C. & Vining, L. C. (1976). Secondary metabolites derived from non-aromatic amino acids. In *The Filamentous Fungi*, vol. II, ed. J. E. Smith & D. R. Berry, p. 478. London: Edward Arnold.

Systematic index

References

Ainsworth, G. C., James, P. W. & Hawksworth, D. L. (1971). *Ainsworth & Bisby's Dictionary of the Fungi.* Kew: Commonwealth Mycological Institute.
Booth, C. (1971). *The Genus Fusarium.* Kew: Commonwealth Mycological Institute.

Subject index